James David Marwick

Charters and Documents Relating to the Collegiate Church and Hospital of the Holy Trinity

James David Marwick

Charters and Documents Relating to the Collegiate Church and Hospital of the Holy Trinity

ISBN/EAN: 9783337235970

Printed in Europe, USA, Canada, Australia, Japan

Cover: Foto ©berggeist007 / pixelio.de

More available books at **www.hansebooks.com**

CHARTERS AND DOCUMENTS

RELATING TO

THE COLLEGIATE CHURCH AND HOSPITAL OF THE HOLY TRINITY,

AND

THE TRINITY HOSPITAL, EDINBURGH.

A.D. 1460–1661.

EDINBURGH:
PRINTED FOR THE SCOTTISH BURGH RECORDS SOCIETY.

MDCCCLXXI.

This collection of Charters was printed primarily for the Corporation of Edinburgh, as administrators of Trinity Hospital, and the Scottish Burgh Records Society were courteously authorised to print [illegible] copies for the use of the Members of the Society.

TABLE OF CONTENTS.

Pages

PREFACE xi

ABSTRACT of Bulls, Charters, Acts of Parliament, and other Documents relating to the Collegiate Church and Hospital of the Holy Trinity, and the Trinity Hospital, Edinburgh xiii

BULLS, CHARTERS, &c.

I. PROMULGATION by the BISHOP of GLASGOW of a BULL by POPE PIUS the SECOND, requiring him to unite and incorporate the Hospital of Soltray with the Church and Hospital of the Holy Trinity of Edinburgh. Bull dated Rome, 23d October 1460, and Promulgation made at Linlithgow 6th March 1461-2. Nos. I. and II. of Abstract 3-15

II. LETTER by QUEEN MARY of GUELDRES, dated Perth, 25th March 1462, craving the Archbishop of St Andrews to confirm the founding and endowing of the Collegiate Church and Hospital of the Holy Trinity, with Ratification and Confirmation thereof by the Archbishop. Dated St Andrews, 1st April 1462. No. III. of Abstract 15-29

III. BULL by POPE PIUS the SECOND confirming the annexation of the Hospital of Soltray to Trinity Hospital. Dated 18th June 1462. No. IV. of Abstract 29-34

IV. BULL by POPE PIUS the SECOND confirming the foundation of the Collegiate Church and Hospital of the Trinity, and the union to the said Hospital of the Hospital of Soltray; as also uniting to Trinity Hospital the Chapel of Utherogall in Fife. Dated at the Abbey of St Salvator, 10th July 1462. No. V. of Abstract 35-39

V. BULL by POPE PIUS the SECOND granting a plenary indulgence to all who should visit the Collegiate Church or Hospital of the Trinity during the

Pages

feast of the dedication of the Church, on the 10th of July or its octaves, for the period of five years, and on the successive recurrence of the day of dedication every fifth year for the period of fifty years. Dated at Tibur, 29th August 1463. No. VI. of Abstract . . . 39–43

VI. TRANSUMPT, dated 21st March 1525, of CHARTER by JAMES, Archbishop of St Andrews, annexing the Church of Dannottar to the Collegiate Church of the Holy Trinity, for the support of two prebendaries in the said College. Dated Edinburgh, 14th November 1502. No. VII. of Abstract 44–54

VII. LETTER by KING JAMES the FIFTH to Pope Clement the Seventh praying his Holiness to grant indulgences to those who, during the lifetime of John Dingwall, then provost of Trinity College, should visit the said College on the Feast of the Holy Trinity, and aid in the completion of the building. Dated Stirling, 22d March 1531. No. IX. of Abstract . 54–55

VIII. CHARTER by QUEEN MARY, under her Great Seal, whereby she granted the kirk-livings to the Provost, Bailies, Councillors and Community of Edinburgh, for the sustentation of the ministry and hospitality within the Burgh. Dated Edinburgh, 13th March 1566–7. No. XI. of Abstract . . 56–63

IX. LETTERS of REMISSION by KING JAMES the SIXTH, under his Great Seal, dispensing with the erection of an hospital on Blackfriars yards, and granting the site and buildings of Trinity College to the Provost, Bailies, Councillors and Community of Edinburgh for the purposes of an hospital. Dated Edinburgh, 3d January 1566–7 [1567–8]. No. XIII. of Abstract 63–66

X. CHARTER by KING JAMES the SIXTH, under his Great Seal, whereby he granted the Collegiate Church of the Holy Trinity, with the churchyard, houses, buildings, orchards, yards, and pertinents of the same to Sir Simon Preston, provost, and his successors the provosts, and to the Bailies, Councillors and Community of Edinburgh, for the building of an hospital thereon, and for the support of the poor who might be placed therein. Dated Edinburgh, 12th November 1567. No. XIV. of Abstract . . . 67–72

XI. CHARTER by KING JAMES the SIXTH, under his Great Seal, known as "The Foundation Charter of the College of Edinburgh," whereby he confirmed Queen Mary's charter of 13th March 1566–7, and of new granted the subjects thereby conveyed, to the Provost, Bailies, Councillors, and Community of Edinburgh, for the sustentation of the ministry, the help of the poor, the repairing of schools, and the increase of letters and science at their dis-

Pages

erection, with power to appoint and dismiss professors, and to provide suitable places for them in which to reside and teach. Stirling, 14th April 1582. No. XVI. of Abstract 73–79

XII. CONTRACT between the Provost, Bailies, Councillors and Deacons of Edinburgh on the one part, and Mr Robert Pont, provost of Trinity College, on the other part, by which the latter agreed to resign the provostry of the said College, with the whole rights, profits, and duties thereof to be applied for the hospitals, college, and schools of Edinburgh, the poor and scholars of the same, and the former agreed to pay Pont three hundred merks sterling and an annual rent of £160 secured over the common mills of the Burgh. Edinburgh, 20th April 1585. No. XVII. of Abstract . . . 80–83

XIII. CHARTER by KING JAMES the SIXTH, under his Great Seal, whereby he granted to the Provost, Bailies, Councillors, and Community of Edinburgh, the provostry of Trinity College, and whole rights, profits, and duties of the same, including the parish churches of Soltray and Lempitlaw, and the place and yards of Dingwall Castle, for the sustentation of the poor within the hospitals, and of poor scholars within the college and schools of the Burgh. Dated Dunfermline, 23d June 1585. No. XVIII. of Abstract . . 83–88

XIV. CHARTER by KING JAMES the SIXTH, under his Great Seal, whereby he confirmed his Charter of 23d June 1585, and, being resolved to alter the destination of the whole fruits, profits, and emoluments of Trinity College, as well those pertaining to the provost as to the prebendaries, chaplains, and other members thereof, and to transfer the same to the use of the ministers, the teaching of literature, and the sustaining of the poor, he of new granted to the Provost, Bailies, Councillors and Community of Edinburgh, the whole property, revenues, and rights which belonged to Trinity College, and to the provost, prebendaries and other members thereof, to be applied at their discretion for the sustentation of the ministers, college, grammar schools, and poor. Dated Holyrood, 26th May 1587. No. XX. of Abstract 89–99

XV. CHARTER by KING JAMES the SIXTH, under his Great Seal, whereby he confirmed, *inter alia* Queen Mary's charter of 13th March 1566–7, and the charters by himself dated 12th November 1567, 23d June 1585, and 26th May 1587, and of new granted to the Provost, Bailies, Councillors and Community of Edinburgh, the whole subjects contained in the charters thereby confirmed, for the support of the ministers and poor, and the upholding of the College lately erected by them. Dated Holyrood, 29th July 1587. No. XXIII. of Abstract 100–105

Pages

XVI. CHARTER by KING JAMES the SIXTH, under his Great Seal, known as "The Golden Charter," whereby he confirmed all charters, infeftments, rights, titles, grants, liberties, and immunities made, granted, or confirmed by him and his predecessors to the Burgh of Edinburgh, and to the churches, colleges, ministers, and hospitals of the same, and specially *inter alia* Queen Mary's charter of 13th March 1566–7, and his own charters, dated respectively 14th April 1582, 23d June 1585, 26th May 1587, and 12th November 1567, with the whole subjects therein contained; and he of new granted to the Provost, Bailies, Councillors and Community of Edinburgh, and their successors, the Burgh of Edinburgh, with the ports and havens of Leith and Newhaven, and the several lands, rights and privileges therein specified; and he united and incorporated with the said Burgh *inter alia* the several church lands and others therein set forth, with the provostry and prebends of Trinity College and Hospital, etc., constituting the whole a free royal burgh. Dated Holyrood, 15th March 1603. No. XXVI. of Abstract 106–125

XVII. CHARTER by KING JAMES the SIXTH, under his Great Seal, whereby he confirmed all infeftments, mortifications, gifts, and dispositions made by him, and by his mother and others his predecessors, in favour of the Burgh of Edinburgh, and of the College, schools, and hospital thereof, for the sustentation of the ministry, and of the masters, regents, and other professors in the College, and of poor scholars and other poor within the Burgh; and of new granted to the Provost, Bailies, Councillors and Community of Edinburgh, the several subjects therein specified for the sustentation of the ministers and of the poor within the Burgh and hospitals thereof, and of poor scholars within the College and schools of the same; and he farther conveyed to the said Provost, Bailies, Councillors, and Community, the Kirk of Field, with the Archdeaconry of Lothian and Church of Currie annexed thereto, and their respective pertinents, for the utility and advantage of the College of the said Burgh. Dated Beauvoir Castle, 10th August 1612. No. XXIX. of Abstract 126–148

XVIII. CHARTER of CONFIRMATION and NOVODAMUS by KING CHARLES the FIRST, under his Great Seal, whereby he confirmed *inter alia* Queen Mary's charter of 13th March 1566–7, and the charters by King James the Sixth, dated 14th April 1582, 26th May 1587, 29th July 1587, and 10th August 1612, excluding any right of regality comprehended in the said charters, and restricting the office of Sheriff and Coroner, and jurisdiction of the same, and the right to hold Guild Courts to the bounds therein described. Dated Newmarket, 23d October 1636. No. XXXIII. of Abstract 149–167

APPENDIX.

Pages

1. ACT OF PARLIAMENT, intituled, "Act in fauoris of the Hospitall of Edinburgh." Passed at Edinburgh on 11th November 1579 [1579, c. 51.] 169

2. ACT OF PARLIAMENT, intituled, "Annexatioun of the temporalities of benefices to the Crown." Passed at Holyrood on 29th July 1587 [1587, c. 8.] 171

3. GENERAL REVOCATION presented by King James the Sixth to Parliament at Holyrood, 29th July 1587 [1587, c. 14.] . . . 175

4. ACT OF PARLIAMENT, intituled, "Ratificatioun of the landis and annuellis mortifiet to the Ministerie and Hospitall of Edinburgh." Passed at Edinburgh on 5th June 1592 [1592, c. 82] 176

5. ACT OF PARLIAMENT, intituled, "Confirmatioun to the Burgh of Edinburgh of thair annuellis." Passed at Edinburgh on 21st July 1593 [1593, c. 41.] 178

6. ACT OF PARLIAMENT, intituled, "Act in fauoris of the Burgh of Edinburgh." Passed at Edinburgh on 11th July 1606 [1606, c. 31.] . . 179

7. WARRANT for an ACT OF PARLIAMENT, superscribed by King James the Sixth. Dated 22d October 1612 181

8. ACT OF PARLIAMENT, intituled, "Ratificatioun of diuers infeftmentis grantit to the toun of Edinburgh for sustentatioun of Colledge, Ministeris, and Hospitallis." Passed at Edinburgh 4th August 1621 [1621, c. 79.] . 182

9. ACT OF PARLIAMENT, intituled, "Ratification of his Majesties new Charter of Confirmation in favors of the Burgh of Edinburgh." Passed at Edinburgh 22d March 1661 [1661, c. 123.] 186

PREFACE.

N attempt has been made to collect in this volume the several Bulls, Charters and other documents which relate to the constitution of the Collegiate Church and Hospital of the Holy Trinity, Edinburgh, and to the gifts by successive Sovereigns to the Magistrates and Town Council of the City, after the Reformation, of the property and revenues of Trinity College and other ecclesiastical establishments for the support of the charity still known as the Trinity Hospital, and for other pious uses. Most of these documents have been previously printed in one form or another, and at different times, to serve temporary purposes, but so inaccurately as to render the meaning obscure.

It is hoped that the present collection will be found more accurate and complete than any hitherto made.

The labour involved in making this collection has been greatly increased by the disappearance of several of the Charters, which seem to have been in the archives of the City at a period not very remote. It would have been satisfactory to have verified the translations of the various charters now given, by a reference

to the Signatures on which the Charters proceeded. Only one of the Signatures, however,—that of the charter of 1612,—has been found. The disappearance of such documents, and the decay of those that remain, is the best possible inducement to the Corporation to persevere in the duty of having all its more important muniments printed ; and it may be hoped that the present collection will be useful in the administration of the important charity which has been entrusted to the Magistrates and Council.

J. D. MARWICK.

10 Bellevue Crescent,
Edinburgh, *10th October* 1870.

ABSTRACT

OF

BULLS, CHARTERS, ACTS OF PARLIAMENT,

AND OTHER DOCUMENTS

RELATING TO

THE COLLEGIATE CHURCH AND HOSPITAL OF THE HOLY TRINITY,

AND TO

THE TRINITY HOSPITAL, EDINBURGH.

[The Asterisks denote those Charters or Documents which have not been found in the Archives of the City, or are deposited elsewhere.]

ABSTRACT, &c.

I. BULL by POPE PIUS THE SECOND, setting forth the erection of the Hospital of Soltray into a dignity named the Chancellorship in the Church of St Andrews, the foundation of the Collegiate Church and Hospital of the Holy Trinity near Edinburgh by Queen Mary of Gueldres, widow of King James the Second, and the desire of the Queen and of her son King James the Third that the Hospital of Soltray should be suppressed as a Chancellorship, and be united and annexed to the Church and Hospital of the Holy Trinity, and requiring the Bishop of Glasgow, after due enquiry, and on obtaining the consent of the parties interested, to suppress the said Chancellorship, and after reducing the Hospital of Soltray to its former state, to unite and incorporate the same with the Church and Hospital of the Holy Trinity. Dated at Rome the 10th day before the Calends of November [23d October] 1460.

Copy in the Archives of the City.
Charters relating to Trinity College and Hospital, Edinburgh, No. I., pp. 4–10.

II. PROMULGATION by ANDREW, Bishop of Glasgow, of the above Bull, which is therein engrossed *verbatim*, setting forth the proceedings had by him under the Papal Commission, suppressing and extinguishing the said Chancellorship, reducing the Hospital of Soltray to its former state, and uniting, annexing and incorporating the same with the College and Hospital of the Holy Trinity. Promulgation made at Linlithgow on 6th March 1461–2.

Transumpt in the Archives of the City.
Charters relating to Trinity College and Hospital, Edinburgh, No. I., pp. 3–15.

*III. LETTERS by QUEEN MARY OF GUELDRES, under her Great Seal, dated at Perth 25th March 1462, whereby, for the causes therein set forth, she craved James Kennedy, Archbishop of St Andrews, to confirm the founding and endowing of the Collegiate Church and Hospital of the Holy Trinity of Edinburgh; with

Ratification and Confirmation by the said Archbishop of the said foundation and endowment. Dated at St Andrews 1st April 1462.

Sir Lewis Stewart's MS. Collections.
Charters of the Collegiate Churches of Midlothian, pp. 63–71.
Charters, &c., relating to Trinity Church and Hospital, Edinburgh, No. II., pp. 15–29.

*IV. BULL by POPE PIUS THE SECOND, addressed to Queen Mary of Gueldres, confirming the annexation of the Hospital of Soltray to the Hospital of the Holy Trinity of Edinburgh. Dated 14 Kal. Julii (18th June) 1462.

Theiner, Vetera Monumenta Hibernorum et Scotorum, No. 818, p. 439.
Charters, &c., relating to Trinity Church and Hospital, Edinburgh, No. III., pp. 29–34.

*V. BULL by POPE PIUS THE SECOND, reciting the foundation by Queen Mary of Gueldres of the Collegiate Church and Hospital of the Holy Trinity, Edinburgh, the erection of the Hospital of Soltray into a Chancellorship of the Church of St Andrews, the suppression of the said Chancellorship, and the union of the Hospital of Soltray to the Trinity Hospital, confirming the said foundation and union, uniting the Chapel of Utherogall, in the County of Fife to Trinity Hospital, and of new erecting Trinity Church into a Collegiate Church, with Collegiate insignia. Dated at the Abbey of St Salvator, in the diocese of Clusium, the 6th Ides of July (10th July) 1462.

Theiner, Vetera Monumenta Hibernorum et Scotorum, No. 821, p. 442.
Munimenta Britannica ex Autographis Romanorum Pontificum deprompta, vol. xxxiv., p. 266 (MS. British Museum, addit. MSS. 15,351 and 15,400 Plut.)
Charters, &c., relating to Trinity Church and Hospital, Edinburgh, No. IV., pp. 35–39.

*VI. BULL by POPE PIUS THE SECOND, granting a plenary indulgence to all who in a devout spirit of contrition should visit the Collegiate Church or Chapel or Hospital of the Holy Trinity, Edinburgh, in the course of five years, during the feast of the dedication of the Church, on the 10th of July, or the following week, called its Octaves, and to those who should have been lawfully prevented from making such visitation as they had wished to do, likewise to the poor faithful in Christ dwelling in the said Church or Hospital for the time, who had died there, and who at the moment of death and at other times were contrite in heart, and confessed with their mouth to the best of their remembrance. This Bull was appointed to come first into effect on the day of dedication in the year next ensuing, and to continue in force on its successive recurrence every fifth year for the period of fifty years only. One-third of the offerings of each person during the Octaves was reserved for the Papal treasury, to assist in carrying on war with the infidels; the other two-thirds were to be applied towards completing the building of the

Church itself. Dated at Tibur the 6th of the Kalends of September (27th August) 1463.

Munimenta Britannica ex Autographis Romanorum Pontificum deprompta, vol. xxxiv., p. 266 (MS. British Museum. addit. MSS. 15,351 and 15,400 Plut.)
Charters of the Collegiate Churches of Midlothian, p. civ.
Charters, &c., relating to Trinity Church and Hospital, Edinburgh, No. V., pp. 39–43.

VII. TRANSUMPT of CHARTER by JAMES, Archbishop of St Andrews, whereby, with advice of the Prior and Chapter of St Andrews, he created and erected two prebendaries to officiate in the Collegiate Church of the Holy Trinity, Edinburgh under the title of Dean of the said Church and Prebendary of Dunnottar respectively, and annexed thereto, and incorporated therewith, the Church of Dunnottar for the support of the said prebendaries. Charter dated Edinburgh 14th November 1502. Transumpt from the Register of the Monastery of St Andrews, made in the Trinity Church of St Andrews, in presence of James Symson, official of St Andrews, on 21st March 1525, under the subscription of Robert Lauson, notary.

Original Transumpt in the Archives of the City.
Charters, &c., relating to Trinity Church and Hospital, Edinburgh, No. VI., pp. 44–54.
Inventory of City Charters, vol. v. p. 248.

*VIII. CONFIRMATION by POPE JULIUS THE SECOND, of the annexation and incorporation of the Church of Dunnottar with the Collegiate Church of the Holy Trinity. Dated in 1504.

Registrum Sancte Trinitatis de Edinburgh, fol. 20.
Charters of the Collegiate Churches of Midlothian, p. 143.

*IX. LETTER by KING JAMES THE FIFTH to POPE CLEMENT THE SEVENTH, praying his Holiness to grant indulgences to those who, during the lifetime of John Dingwall, then provost of Trinity College, should visit the said College on the feast of the Holy Trinity, and aid in the completion of the building. Dated at Stirling the 22d day of March 1531.

Theiner, Vetera Monumenta Hibernorum et Scotorum, No. 1025, p. 597.
Charters, &c., relating to Trinity Church and Hospital, Edinburgh, No. VII., pp. 54–55.

*X. CHARTER by QUEEN MARY, under her Great Seal, whereby she gave, granted, and disponed to the Provost, Bailies, Councillors and Community of the Burgh of Edinburgh, and their successors, the place and yards which belonged to the Black-friars, formerly called the Preaching Friars, with the cemeteries and other pertinents

of the same, for the construction and erection of an Hospital thereon for the relief and assistance of the poor; which Hospital was appointed to be commenced within a year from the date of the infeftment on the Charter, and to be finished within ten years next thereafter. Dated at St Andrews 16th March 1562–3.

This Charter is described in the Letters of Remission, No. XIII., but has not been discovered in the Archives of the City. Neither is it recorded in the Registers of the Great Seal or of the Privy Seal.

XI. CHARTER by QUEEN MARY, under her Great Seal, whereby with the advice of the Lords of her Privy Council, she gave, granted, disponed, and for ever confirmed to the Provost, Bailies, Councillors and Community of the Burgh of Edinburgh, for the sustentation of the ministry and hospitality within the said Burgh, all lands, tenements, houses, buildings, churches, chapels, yards, orchards, crofts, annual rents, fruits, duties, profits, emoluments, rents, alms, daill-silver, obits, and anniversaries whatsoever which pertained of before to any chaplainry, altarage, and prebend founded in any church, chapel, or college within the liberty of the said Burgh, with the manor, places, orchards, lands, annual rents, emoluments, and duties whatsoever which formerly belonged to the Dominican or Preaching Friars, and to the Minorite or Franciscan Friars of the said Burgh; together with all lands, houses, and tenements lying within the said Burgh and liberty thereof, with all annual rents leviable from any house, land, or tenement within the said Burgh, given and doted to any chaplainries, altarages, churches, burials, or anniversaries within the kingdom, or owing furth of the said Burgh or tenements thereof to whatsoever other benefice or chaplainry. And she united and incorporated the subjects thereby conveyed, into one body, to be called in all time coming the Foundation of the Ministry and Hospitality of Edinburgh, and declared that one sasine, taken at the Tolbooth of Edinburgh, should be a sufficient sasine in all time thereafter. Dated at Edinburgh 13th March 1566–7.

Original Charter in the Archives of the City.
Charters, &c., relating to Trinity Church and Hospital, Edinburgh, No. VIII., pp. 56–63.
Inventory of City Charters, vol. iii. p. 1.

XII. INSTRUMENT of SASINE taken in the upper Tolbooth of the Burgh of Edinburgh by William Stewart, notary, proceeding upon the said charter and precept therein. Dated 29th March 1567.

Original Instrument in the Archives of the City.
Inventory of City Charters, vol. iii. p. 5.

*XIII. LETTERS of REMISSION by KING JAMES THE SIXTH, under his Great Seal, narrating the Charter by Queen Mary of 16th March 1562–3 [No. X.], dispensing with the erection of an Hospital on the Blackfriars' Yards, and granting the

site and buildings, etc., of Trinity College to the Provost, Bailies, Councillors and Community of the Burgh of Edinburgh for the purposes of an Hospital. Dated at Edinburgh 3d January 1566–7 [1567–8], and of the King's reign the first.

This Charter has not been discovered in the Archives of the City, nor is it recorded in the Registers of the Great Seal or Privy Seal. The print here given is taken from a copy engrossed without much accuracy in a Memorial to Counsel in 1833. The date is obviously erroneous, as the King's reign commenced on 29th July 1567.

Charters, &c., relating to Trinity Church and Hospital, Edinburgh, No. IX., pp. 63–66.

*XIV. CHARTER by KING JAMES THE SIXTH, under his Great Seal, whereby with the advice and consent of the Lords of his Privy Council, he gave, granted, and disponed to Sir Simon Preston, provost of the Burgh of Edinburgh, and his successors the Provosts, and to the Bailies, Councillors and Community of the said Burgh for the time being, the church called the Collegiate Church of the Trinity, with the churchyard, houses, buildings, orchards, yards, and pertinents thereof, formerly occupied and possessed by the provost and prebendaries of the said Collegiate Church, together with the place and part, with the buildings and yards of Trinity Hospital lying contiguous to the Church, with the yard lying on the west side thereof, for the building and construction of an Hospital thereon for the maintenance of such honest poor and impotent persons, aged and advanced in years, or sick, indwellers and inhabitants within the said Burgh, and also for such other old, indigent, and impotent people as should be found fit for receiving such benefits and charity in the said hospital so to be founded; but always without prejudice to the rights of the provost and prebendaries of Trinity College to present so many of the poor, commonly called beidmen, as were then placed and endowed in Trinity Hospital in terms of the old foundation. Dated Edinburgh 12th November 1567.

Registrum Magni Sigilli, Book xxxii., No. 613.
Charters, &c., relating to Trinity Church and Hospital, Edinburgh, No. X., pp. 67–72.

*XV. ACT of PARLIAMENT in favour of the Hospital of Edinburgh, whereby, after reciting Queen Mary's Charter of 13th March 1566–7 [No. XI.], and setting forth that there were chaplainries founded in certain places outwith the Burgh of Edinburgh, of which the Provost, Bailies, and Council were undoubted patrons, but that the service of these chaplainries had altogether ceased, the Provost, Bailies, and Council, and their successors, were empowered to uplift the profits and duties of the said benefices, and to bestow the same to the sustentation of the Ministry and Hospitality of the Burgh. License was farther given to all persons who, "pitying the miserable estate of the poor, and delighting in that good work of erection of an Hospital within the said Burgh," might be disposed to supply the said Hospital with their alms, and support of annual rents, lands, and tenements

lying within the Burgh, to be annexed thereto, for the entertainment of the poor, weak, aged, and sick persons to be sustained therein. Passed at Edinburgh on 11th November 1579.

Acts of the Parliaments of Scotland, vol. iii., p. 169, c. 51.
Charters, &c., relating to Trinity Church and Hospital, Edinburgh, App. No. I., p. 169.

XVI. CHARTER by KING JAMES THE SIXTH, under his Great Seal, whereby with the advice of the Lords of his Privy Council, he confirmed Queen Mary's Charter of 13th March 1566–7 [No. XI.], and of new gave, granted, and disponed to the Provost, Bailies, Councillors and Community of the Burgh of Edinburgh, and their successors, the several subjects conveyed by that charter, to be by them applied in all time coming to the sustentation of the ministry, the help of the poor, the repairing of schools, the increase of letters and science, at the discretion of them and their successors as should seem to them most advantageous. They were also empowered to accept of whatsoever other annual rents and yearly profits, as well without as within the Burgh, might be given by any persons for the maintenance of the ministers of the gospel, the help of the poor, and the sustentation of schools for the increase of science and learning. The King further ratified and confirmed the renunciation and demission made by John Gibb in favour of the said Provost, Bailies, Councillors and Community, for themselves and their successors, and in name and on behalf of the ministers and poor, of all right and title which Gibb had to the provostry of the Kirk of Field. The said Provost, Councillors, and their successors were further empowered to build and repair sufficient houses and places for the reception, habitation, and teaching of the professors of the schools of grammar, humanity, and the languages, philosophy, theology, medicine, and law, or any other liberal sciences, and these several purposes it was declared should be held to be no invasion of the original mortification. By this Charter also, the Provost, Bailies, and Councillors, and their successors, with advice of their ministers, were empowered to choose the professors of the College of Edinburgh, and also to place and remove them as should be expedient: and all others were discharged from professing or teaching the said sciences within the liberty of the Burgh unless they were permitted to do so by the Town Council; but it was provided that the ministers present and to come, serving in the churches thereby conveyed, should be sustained out of the readiest fruits of the same. Dated at Stirling 14th April 1582.

Original Charter in the Archives of the City.
Charters, &c., relating to Trinity Church and Hospital, Edinburgh, No. XI., pp. 73–79.
Inventory of City Charters, vol. iii. p. 12.

XVII. CONTRACT between the Provost, Bailies, Councillors and Deacons of the Burgh of Edinburgh, for them and their successors, on the one part, and Mr Robert Pont, provost of Trinity College, on the other part. By this contract the said

Robert Pont demitted, renounced, and resigned into the hands of the King, the benefice of the Trinity College, with all and sundry churches, teinds, glebes, buildings, yards, annual rents, advocation, donation, and right of patronage of prebendaries, chaplainries, and donation of beidmenships, bedlyars, and other officers pertaining to the said provostry and Hospital, with the parish kirk, parsonage, and vicarage of Soltray and Lempitlaw, and with the place, orchard, and yard called Dingwall Castle, and all and sundry other fruits, emoluments, rights, casualties, profits, and duties belonging to the said provostry, in favour of the Provost, Bailies, Councillors and Community of Edinburgh, and their successors, to remain with them perpetually in all time coming in pure and perpetual alms. To be applied and disponed upon by them to the maintenance, help, and support of their said Hospitals, College, and Schools, the poor and scholars of the same, as they should think expedient, and as they should answer to God at the latter day. He further undertook to deliver to them the writs, evidents, and rentals of the Provostry College, and Hospital, and consented that they should enter into immediate possession thereof, with power to appoint all officers needful for managing the same, and to uplift the revenues thereof. On the other hand, the said Provost, Bailies, and Councillors paid the said Robert Pont three hundred merks sterling "in contentation of all grassums, entry silver, and other casualties which he might have received of the said benefice during his lifetime;" and they further agreed to pay him yearly during his life £160 Scots at Whitsunday and Martinmas by equal portions, beginning the first term's payment at Martinmas 1585; which annuity they engaged to secure by infefting him in an annual rent of corresponding amount out of the common mills of the Burgh. The Contract contains a clause of warrandice by Pont that the yearly rent of the Provostry was fully worth the said sum of £160 Scots. Dated at Edinburgh 26th April 1585.

Original in the Archives of the City.
Charters, &c., relating to Trinity Church and Hospital, Edinburgh, No. XII., pp. 80–83.
Inventory of City Charters, vol. v. p. 161.

*XVIII. CHARTER by KING JAMES THE SIXTH, under his Great Seal, whereby, with the advice of the Lords of his Privy Council, he granted to the Provost, Bailies, and Councillors of the Burgh of Edinburgh, and their successors, the benefice of the Provostry of the Collegiate Church of the Trinity, with the churches, teinds, glebes, manses, buildings, orchards, yards, annual rents, advocations, donations, and the right of patronage of prebendaries and chaplainries, and donations of poor oratours, commonly called beidmen and bedlyaris, and other officers of the said provostry and Hospital of Trinity College, founded near the same; with the parish churches of Soltray and Lempitlaw, and other churches and teinds annexed to the said provostry, with the place, orchard, and yard called Dingwall Castle, and all other fruits, emoluments, rights, casualties, profits, and duties belonging to the said provostry. To be intromitted with and disponed upon by the said Provost,

Bailies, and Councillors, and their successors in all time coming, as they would answer to Almighty God at the last day, for the sustentation of the aged, decrepit, orphans, and poor within the Hospitals, and of poor scholars within the College and Schools founded by them within the Burgh, as they should answer to God in the last judgment. To be holden to the effect foresaid in pure and perpetual alms, under burden of sustaining the ministers serving the cure of the churches belonging to the said provostry, and their successors, or of paying the third part of the fruits of the provostry for their sustentation, at their option and choice. Dated at Dunfermline 23d June 1585, in the eighteenth year of the King's reign.

Registrum Magni Sigilli, xxxvi., No. 360.
Charters, &c., relating to Trinity Church and Hospital, Edinburgh, No. XIII., pp. 83–88.
Inventory of City Charters, vol. v. p. 164.

*XIX. INSTRUMENT of SASINE, taken at the Kirk of the said College, in favour of the said Provost, Bailies, Councillors, and Community, under the hands of Mr David Guthrie, notary, and Mr Alexander Guthrie, notary and town clerk of Edinburgh. Dated at Edinburgh 13th August 1585

Original Instrument in the Archives of the City.
Inventory of City Charters, vol. v. p. 164.

*XX. CHARTER by KING JAMES THE SIXTH, under his Great Seal, whereby, with the advice and consent of the Lords of his Privy Council, he confirmed his Charter of the provostry of Trinity College dated 23d June 1585 [No. XVIII.], and farther, after mature deliberation, being fully resolved to alter the destination of the whole fruits, profits, and emoluments of Trinity College, as well those pertaining to the provost as to the prebendaries, chaplains, and other members thereof, the services for which these were formerly founded being nowise necessary, and to transfer the same to the use of the ministers, the teaching of literature, and the sustaining of the poor, he, with advice foresaid, of new granted to the Provost, Bailies, Councillors, and Community of the Burgh of Edinburgh, and their successors, the provostry of Trinity College and whole pertinen's thereof, including the donation and right of patronage of prebends and chaplainries of the said College, with the donation of beidmen and bedlyars, and of other officers of the said provostry and Hospital of the said College founded near the same, together with the churches of Soltray and Lempitlaw, &c., with the place, orchard, and yard of Dingwall Castle, &c., and all other pertinents belonging to the said provostry, and also the whole properties and revenues belonging to, or which were formerly possessed by, the prebendaries and chaplains of the said College, with all the properties and revenues founded and mortified to the said College, provost, prebendaries, and members thereof, or to the upholding of the church houses and buildings of the said College;

with power to uplift the whole profits and produce of the subjects conveyed, and to apply the same to the sustaining of the Ministers, College, Grammar Schools, and Poor people, at their own good discretion, whereanent the King burdened their consciences. And the grantees were relieved from the obligations imposed by any clauses in the foundation of Trinity College to present any prebendary or chaplain to the prebends or chaplainries then vacant, or that might thereafter become vacant, or to grant to them any special title to the same; which clauses were thereby annulled and abrogated, that the mortification contained in the Charter might receive effect, and that the foresaid profits might be all collected and ingathered together in one rental and disponed to the foresaid uses. Moreover, the King willed and granted that the said Provost, Bailies, Councillors, and Community, and their successors, should have the full right of property in all the subjects thereby conveyed, with the superiority of the whole lands, and all the rights incident thereto, as fully as the provost, prebendaries, and hospitallers could have enjoyed or exercised the same, by reason of their foundation or otherwise. Farther, the said Provost, Bailies, Councillors, and Community, and their successors, were empowered to sustain within the Hospital lately built and repaired by them within a part of Trinity College Church, as many poor persons as could be conveniently sustained upon the rents of the Hospital of Trinity College, and to apply Trinity Hospital, which was then in a ruinous condition, to whatever other profitable use should seem most expedient. To be holden in pure and perpetual alms for ever, as in the Charter of 23d June 1585, the grantees being always bound to lay out and expend the whole annual produce of the subjects conveyed to the foresaid uses, and to account therefor to the King and his successors whenever required. But the liferent rights of the prebendaries of the said College then living were expressly reserved. Dated at Holyrood 26th May 1587, in the twentieth year of the King's reign.

Registrum Magni Sigilli, Lib. xxxvi., No. 360.
Charters, &c., relating to Trinity Church and Hospital, Edinburgh, No. XIV., pp. 89-99.
Inventory of City Charters, vol. v. p. 168.

*XXI. ACT of PARLIAMENT intituled, "Annexation of the temporalities of benefices to the crown," whereby King James the Sixth and the three estates of Parliament united, annexed, and incorporated to the crown, to remain therewith in all time coming, all lands, lordships, baronies, etc., and all other commodities, profits, and emoluments whatsoever, as well to burgh as to land, which, at the date of the Act (29th July 1587), belonged to any ecclesiastical or beneficed person of whatsoever degree, or to any abbey, convent, cloister, friars, nuns, monks, or canons, or to any college-kirk founded for singing, or to any prebendary or chaplainry within the realm, etc. The execution of the act in the levying of the profits was appointed to take effect as at the term of Martinmas 1587. From the annexation there was excepted a variety of subjects therein enumerated, and specially all lands and other

subjects granted for the entertainment of masters and students in colleges and grammar schools, and for sustentation of ministers residing in Burghs where there was no other stipend appointed to them; as also all lands and other subjects granted by the King and by his predecessors before the date of the act, or by any other persons, to any hospital or maisondieu within the realm, and that in favour of the poor and needy, providing the same were not disponed to any other use. Passed at Holyrood on 29th July 1587.

Acts of the Parliaments of Scotland, vol. iii., p. 431, c. 8.
Charters, &c., relating to Trinity Church and Hospital, Edinburgh, App. No. II., p. 171.

*XXII. GENERAL REVOCATION, presented by KING JAMES THE SIXTH to Parliament, whereby, having attained the perfect age of twenty-one years, he revoked all infeftments, gifts, and dispositions set, given, and granted by him in his minority to any persons in fee, feu-farm, or liferent of any hospitals, maisondieus, lands, or rents appertaining thereto, in hurt and prejudice of his conscience, to the end that the said hospitals might be reduced to their first institution for upholding of the poor, excepting from that revocation the rents of the hospital of the Trinity College consigned and given to the new hospital erected by the Provost, Bailies, and Council. And he farther revoked all infeftments made by him in his minority, or by his governors and regents in his name, of any kirk's lands, friar's lands, men's lands, or common lands which anyways fell to the crown, except the infeftments made by Queen Mary and him for the erection and sustentation of hospitals and ministers within burghs where there is no assignation or stipend allowed furth of the thirds of benefices for sustentation of the ministers thereof. Passed at Holyrood, 29th July 1587.

Acts of the Parliament of Scotland, vol. iii. pp. 439-442, c. 14.
Charters, &c., relating to Trinity Church and Hospital, Edinburgh, App. No. III., p. 175.

*XXIII. CHARTER by KING JAMES THE SIXTH, under his Great Seal, to the Provost, Bailies, Councillors, and Community of the Burgh of Edinburgh, whereby he confirmed for ever to them, *inter alia*, Queen Mary's Charter of 13th March 1563[1] [No. XI.], and the Charters by himself dated respectively 12th November 1567 [No. XIV.], 23d June 1585 [No. XVIII.], and 26th May 1587 [No. XX.]. Moreover he of new gave, disponed, and mortified to the said Provost, Bailies, Councillors, and Community, and their successors, for the support of the Ministers and Poor, and for the upholding of the College by them lately erected, all lands, rents, teinds, and other profits and emoluments particularly contained in the charters, &c., so confirmed, to remain with them for ever for the uses therein specified, and not otherwise, according to the form and tenor of the same. But it is provided that the grantees and their successors should be held bound to support the ministers in their churches then serving there, and similar qualified persons who should be ordained

[1] Should be 1566-7.

to serve in the said cures for ever, according to the tenor of the donations and mortifications formerly made to that effect. Dated at Holyrood, 29th July 1587.

Registrum Magni Sigilli, Lib. xxxvi., No. 534.
Charters, &c., relating to Trinity Church and Hospital, Edinburgh, No. XV., pp. 100-105.

*XXIV. ACT of PARLIAMENT, intituled, "Ratification of the landis and annuallis mortifiet to the ministrie and hospitall of Edinburgh," whereby King James the Sixth, with the advice of his estates in Parliament, ratified and approved the donations and mortifications made by Queen Mary and by himself, at divers times, of the lands, benefices, and rents given for sustentation of the Ministry within the Burgh of Edinburgh, and for entertaining the Hospitals thereof; and with the advice of the said estates, he of new annexed the said lands, benefices, and rents to the community of the said Burgh and their successors in favour of their Ministry and Hospital, and ordained a new infeftment to be expede thereupon for their security if it were thought expedient. And he dissolved his general annexation, in so far as it might appear to be extended to any of the premises, or to the annexation previously made in favour *inter alia* of the lands, annual rents, houses, yards, and biggings of the Trinity College, pertaining as well to the provost as to the prebendaries thereof, and common lands and annual rents of the same, which annexation the King, with advice of his said estates in parliament, duly ratified and approved. And he farther annexed other ecclesiastical properties, to remain with the Provost, Bailies, Councillors, and Community of Edinburgh, and their successors in time coming, for sustentation of the said Ministry and Hospital, and declared that none of the subjects previously disponed and newly annexed for sustentation of the said Ministry and Hospital "ever ar or salbe euir comprehendit in the generall annexatioun of the ecclesiasticall landis and rentis to the croun;" and the grantees were declared to have the full right of property and superiority of the foresaid lands, &c., notwithstanding any act or constitution preceding the date of said Act of Parliament. Dated at Edinburgh on 5th June 1592.

Acts of the Parliaments of Scotland, vol. iii., p. 582, c. 82.
Charters, &c., relating to Trinity Church and Hospital, Edinburgh, App. No. IV., p. 176.

*XXV. ACT of PARLIAMENT intituled, "Confirmation to the Burgh of Edinburgh of their Annuallis," whereby the King, with the advice of his estates in Parliament, ratified and approved the act of Parliament passed on 5th June 1592 [No. XXIV.], and revoked and rescinded all and sundry infeftments, gifts, and dispositions made by him to any persons of the lands and other subjects mentioned in the said act since the dates of the respective infeftments, gifts, and dispositions made thereof to the Provost, Bailies, Councillors and Community of the Burgh for the sustentation of the said Ministry, Hospital, and College. Passed at Edinburgh on 21st July 1593.

Acts of the Parliaments of Scotland, vol. iv., p. 31, c. 41.
Charters, &c., relating to Trinity Church and Hospital, Edinburgh, App. No. V., p. 178.

XXVI. CHARTER by KING JAMES THE SIXTH, under his Great Seal, known as the "Golden Charter," in favour of the Provost, Bailies, Councillors, and Community of the Burgh of Edinburgh, by which, for the reasons therein set forth, he, with the advice and consent of the Lords of his Privy Council, as also with the express advice and consent of his treasurer and comptroller, and of his collector general and treasurer of the new augmentations of the church lands of the kingdom to the crown, ratified, approved, and for ever confirmed all charters, infeftments, rights, titles, gifts, grants, liberties, and immunities made, granted, or confirmed by him and his predecessors to the said Burgh, and to the Churches, Colleges, Ministers, and Hospitals of the same, and specially *inter alia* Queen Mary's Charter of 13th March 1566–7 [No. XI.], and the Charters by himself dated respectively 14th April 1582 [No. XVI.], 23d June 1585 [No. XVIII.], 26th May 1587 [No. XX.], 12th November 1567 [No. XIV.], with the whole subjects and others contained in the said Charters, &c. He further, with advice and consent foresaid, conveyed of new to the Provost, Bailies, Councillors, and Community, and their successors, the Burgh of Edinburgh, with the ports and havens of Leith and Newhaven, and the several lands, rights, and privileges therein specified, and he united and incorporated with the said Burgh *inter alia* the several church lands and others therein set forth, with the Provostry and Prebends of Trinity College and Hospital, etc., constituting the whole a free Burgh royal to be held in feu-farm and free burgage for ever. Dated at Holyrood 15th March 1603, in the thirty-sixth year of the King's reign.

Original Charter in the Archives of the City.
Registrum Magni Sigilli, Lib. XLIV., No. 363.
Charters, &c., relating to Trinity Church and Hospital, Edinburgh, No. XVI., pp. 106-125.
Inventory of City Charters, vol. i. p. 679.

XXVII. INSTRUMENT of SASINE under the hand of Mr Alexander Guthrie, town clerk of Edinburgh and notary, proceeding upon the said precept. This instrument bears that Sir John Arnot of Berswick, knight, provost; William Speir, John Jackson, bailies; Richard Dobie, dean of guild; Thomas Speir, treasurer, all of the Burgh of Edinburgh, and Andrew Kneeland, sheriff of Edinburgh, specially appointed by the said Precept passed to the Market Cross, and there the said Sheriff gave to the said Provost for himself, and in name of the Bailies, Councillors, and Community of the Burgh, sasine of the subjects and others contained in the said charter. Dated 5th October 1611.

Original in the Archives of the City.
Inventory of City Charters, vol. i. p. 733.

•XXVIII. ACT of PARLIAMENT intituled "Act in fauores of the Burgh of Edinburgh," whereby King James the Sixth and his estates of Parliament, ratified, approved, and confirmed to the Provost, Bailies, Councillors and Community of the Burgh all

and sundry gifts, mortifications, and infeftments made by his mother and by himself to them of all lands, annual rents, tenements, teinds, provostries, prebendaries, altarages, and other benefices, teinds, rents, and emoluments, with all acts of Parliament made in favour of them, and of the Ministry, College, and Hospitals within the Burgh, subject to the provision that a reasonable and sufficient stipend should be modified to the minister of Currie and his successors. Passed at Edinburgh on 11th July 1606.

Acts of the Parliament of Scotland, vol. iv. p. 303, c. 31.
Charters, &c., relating to Trinity Church and Hospital, Edinburgh, App. No. VI., p. 179.

XXIX. CHARTER by KING JAMES THE SIXTH, under his Great Seal, in favour of the Provost, Bailies, Councillors and Community of the Burgh of Edinburgh, whereby with the advice and consent of the Lords of his Privy Council of his kingdom of Scotland, his Commissioners representing the Officers of State of the said kingdom, he confirmed to them all and sundry infeftments, mortifications, gifts, and dispositions whatsoever made by him, and by his mother and others his predecessors, in favour of them, the College, Schools, and Hospital of the said Burgh, for the sustentation of the Ministry serving the cure at the churches of the said Burgh, of the master regents and other professors serving the cure of the foresaid College and Schools, of poor scholars and other poor, aged, decrepit, indigent persons, orphans and children deprived of their parents, within the said Burgh, of all and sundry benefices, lands, tenements, annual rents, teinds, fruits, rents, and other emoluments therein expressed Moreover the king, with advice and consent foresaid, of new gave, granted, disponed, mortified, and perpetually confirmed to the said Provost, Bailies, Councillors, and Community, and their successors, all and sundry lands and other subjects described in the several Charters thereby confirmed, likewise the provostry of Trinity College, and whole prebends pertaining and belonging to the same, with the churches, &c., of Soltray and Lempitlaw annexed to the said provostry, and the churches, &c., of Ormiston, Kirkurd, and Wemyss annexed of old to the Church of Soltray, the place, &c., called Dingwall Castle, and whole other pertinents of the said provostry; as also the provostry of St Giles' Church, with the churches of Dumbarnie, Potie, and Moncreif, &c., pertaining thereto; the Nunnery of the Schiennes, and the Hospital of St Paul's Work, at the foot of Leith Wynd, with the respective pertinents, to be intromitted with, uplifted, used, and disponed upon in all time coming by the said Provost, Bailies, Councillors and Community and their successors, for the sustentation of the ministers serving the cure at the churches of the said Burgh, and of the aged, decrepid, orphans, and poor within the said Burgh and Hospitals thereof, and poor scholars within the College and Schools of the same. The King further conveyed the Kirk of Field with the Archdeaconery of Lothian and church of Currie annexed thereto, and their respective pertinents, to the said Provost, Bailies, Councillors and Community and their successors, to be intromitted with, used, and disponed upon by them, in all time coming, for the utility and

advantage of the College of the said Burgh, masters, regents, and other professors doing duty within the same. And the whole subjects of the grant are thereby incorporated into one Body, to be called "The foundation of the Ministry and Hospitality of Edinburgh"; and one sasine, to be taken at the Tolbooth of the Burgh, is thereby appointed to be a sufficient sasine in all time coming. To be holden in perpetual alms for ever. Dated at Beavoir Castle 10th August 1612, in the forty-sixth and tenth years of the King's reign respectively.

Original Charter in the Archives of the City.
Registrum Magni Sigilli, Lib. xlvii. No. 34.
Charters, &c., relating to Trinity Church and Hospital, Edinburgh, No. XVII., pp. 126-148.

XXX. INSTRUMENT of SASINE, under the hand of Mr Alexander Guthrie, town clerk of Edinburgh, notary, following upon the said Precept. The infeftment bears to have been taken within the Laigh Tolbooth of the Burgh of Edinburgh, and the sasine to have been given to Sir John Arnot of Berswick, knight, Lord Provost of Edinburgh, for himself, and in name of the Provost, Bailies, Councillors, and Community of the said Burgh. Dated 3d October 1612.

Original Instrument in the Archives of the City.
Inventory of City Charters, vol. III., p. 63.

*XXXI. WARRANT for an ACT of PARLIAMENT superscribed by KING JAMES THE SIXTH, in which his Majesty and his estates of Parliament are stated to have ratified and approved the Charter dated 10th (erroneously stated 7th) August 1612 [No. XXIX.]. And the Lord Clerk Register and his deputes are ordained to extend an Act of Parliament thereupon, and to engross the Charter thereby confirmed in more ample form. Dated 22d October 1612.

Inventory of City Charters, vol. iii. p. 63.
This Warrant has not been found in the Archives of the City.
Charters, &c., relating to Trinity Church and Hospital, Edinburgh, App. No. VII., p. 181.

*XXXII. ACT of PARLIAMENT intituled "Ratificatioun of divers infeftmentis grantit to the Toun of Edinburgh for sustentatioun of Colledge Ministeris and Hospitallis," whereby King James the Sixth, with advice of the estates of Parliament, ratified and approved the several charters granted by him to the Provost, Bailies, Councillors, and Community of the Burgh of Edinburgh and their successors, and *inter alia*, those of the following dates, viz., 14th April 1582 [No. XVI.], 26th May 1587 [No. XX.], 29th July 1587 [No. XXII.], 7th [10th] August 1612 [No. XXIX.], and further ratified and approved the erection of the College for profession of theology, philosophy, and humanity, together with the foresaid mortifications made by him, either to the use of the College or to the use of the Ministry and Hospital of the burgh. Moreover,

the King ordained the College of Edinburgh to be called in all time coming "King James' College," and disponed to the grantees and their successors, in favour of the said Burgh and of the said College, and of the rectors, regents, bursars, and students within the same, all liberties, freedoms, immunities, and privileges appertaining to a free college, and that as amply as any college had or enjoyed within the realm. Passed at Edinburgh 4th August 1621.

Acts of the Parliament of Scotland, vol. iv. p. 670, c. 79.
Charters, &c., relating to Trinity Church and Hospital, Edinburgh, App. No. VIII., p. 182.

XXXIII. CHARTER of CONFIRMATION and NOVODAMUS granted by KING CHARLES THE FIRST, under his Great Seal, in favour of the Provost, Bailies, Councillors and Community of the Burgh of Edinburgh, whereby, on the narrative of the Charter of King James the Sixth, dated 15th March 1603 [No. XXVI.], and of a renunciation and resignation by the said Provost, Bailies, Councillors, and Community of certain rights conferred, or alleged to be conferred upon them by the said Charter, and of a petition by them to his Majesty to ratify the said Charter, and all their ancient infeftments therein contained, the King ratified and approved *inter alia* Queen Mary's Charter, dated 13th March 1566–7 [No. XI.], and the Charters of King James the Sixth of the following dates, viz., 14th April 1582 [No. XVI.], 26th May 1587 [No. XX.], 29th July 1587 [No. XXIII.], 7th [10th] August 1612 [No. XXIX.], providing always that the said Confirmation should not be extended to any right of regality comprehended in the said charters, nor should be farther extended to the offices of Sheriff and Coroner, and jurisdiction of the same, and to the holding of Guild Courts beyond the bounds therein described, nor to other matters therein set forth. Farther, the said Charter *inter alia* conferred upon the grantees the patronage of all the churches within the Burgh, and united and incorporated the whole subjects therein described into a royal city, with all the liberties, privileges, and immunities belonging to a city or royal burgh. Dated at Newmarket 23d October 1636, in the twelfth year of the King's reign.

Original in the Archives of the City.
Registrum Magni Sigilli, Lib. LV., No. 82.
Charters, &c., relating to Trinity Church and Hospital, Edinburgh, No. XVIII., pp. 149-167.
Inventory of City Charters, vol. i. p. 32.

*XXXIV. INSTRUMENT of SASINE following thereon, under the hand of Mr Francis Hay and Mr Peter Alesso, notaries, dated 6th February, and recorded in the Particular Register of Sasines at Edinburgh, 13th March 1637.

Original Instrument in the Archives of the City.
Inventory of City Charters, vol. i. p. 832.

*XXXV. SIGNATURE granted and supersigned by KING CHARLES THE SECOND, ordaining a Charter to be passed under the Great Seal ratifying, approving, and confirming in favour of the Provost, Bailies, Councillors and Community of the Burgh of Edinburgh and their successors, all charters, gifts, grants, and other writs and evidents made by any of the Kings or Queens of Scotland, Governors or Regents thereof for the time, or by their Commissioners, or by any other persons to the Burgh, or to the Kirks, College, Ministers, and Hospitals of the Burgh, or to the Magistrates, Councillors, Burgesses and Commonalty of the same, together with the whole erections, liberties, rents, lands, tenements, jurisdictions, superiorities, mortifications, patronages and others pertaining and disponed to them. Dated at Whitehall 10th September 1660.

This Signature has not been found in the Archives of the City.

*XXXVI. ACT of PARLIAMENT intituled "Ratification of his Majesties new Charter of Confirmation in favour of the Burgh of Edinburgh," whereby King Charles the Second, with the advice and consent of his estates of Parliament, ratified, approved, and confirmed the Signature dated 10th September 1660 [No XXXV.], and willed and declared that the said act should be as valid and sufficient as if the said signature, and the charter to follow thereupon, were already extended and passed under the Great Seal. It is also decerned and ordained that the foresaid signature, and the charter to pass thereupon, with the confirmation thereof contained in the said Act of Parliament, should be a good and perfect right to the said Provost, Bailies, Councillors, and Commonalty and their successors, for bruiking and joysing conform to their rights the whole lands and others granted by King Charles the First, or any others his royal predecessors, to the said Burgh, or to the Kirks, Colleges, Ministers, and Hospitals thereof; and the Lord Treasurer and Lords of Exchequer were ordained to pass to the said Burgh and Colleges and Hospitals thereof the particular infeftments and grants foresaid. Passed at Edinburgh 22d March 1661.

Acts of the Parliament of Scotland, vol. vii. p. 84.
Charters, &c., relating to Trinity Church and Hospital, Edinburgh, App. No. IX., p. 186.

CHARTERS AND DOCUMENTS

RELATING TO

THE COLLEGIATE CHURCH AND HOSPITAL OF THE HOLY TRINITY, EDINBURGH,

AND

THE TRINITY HOSPITAL, EDINBURGH.

I.

PROMULGATION by Andrew Bishop of Glasgow, on 6th March 1461–2, of the Bull of Pope Pius the Second dated 23d October 1460, authorising the annexation of the Hospital of Soltray to the Collegiate Church and Hospital of the Holy Trinity.

ANDREAS Dei [et] apostolice sedis gratia episcopus Glasguensis, iudex et executor vnicus ad infrascripta auctoritate apostolica specialiter deputatus, Vniuersis Christi fidelibus ad quorum noticias presentes nostre litere peruenerint, salutem. Sciatis nos quasdam literas apostolicas sanctissimi in Christo patris et domini nostri domini Pii diuina fauente clemencia Pape secundi eius vera bulla plumbea cum cordula canapis more Romane curie impendente sigillatas, non viciatas non cancellatas nec in aliqua sui parte suspectas, sed omni prorsus vicio et suspicione carentes, super suppressione et extinctione nominis et dignitatis cancellarie Sanctiandree alias erectorum, et reduccione hospitalis pauperum de Soltre dicte Sanctiandree diocesis in statum pristinum, illoque reducto

ANDREW, by the grace of God and of the Apostolic See, Bishop of Glasgow, sole judge and executor in the matters underwritten specially deputed by apostolic authority, to all the faithful in Christ to whose notice our present letters shall come, greeting: Know that we have with becoming reverence received certain Letters Apostolic, of the most holy father in Christ, and our Lord the Lord Pius the Second, by favour of the divine clemency Pope, sealed with his true leaden bull attached with a hempen cord, after the manner of the Roman court, not vitiated, cancelled or suspected in any part, but absolutely free from all vitiation and suspicion, concerning the suppression and extinction of the name and dignity of the Chancellorship of St Andrews otherwise erected, and the reduction of the Hospital for the poor at Soltray in the said diocese of St Andrews into its former state,

vnione anneccione et incorporacione illius Collegio et Hospitali Sancte Trinitatis nouiter per serenissimam principissam et dominam dominam Mariam Reginam Scocie iuxta Burgum de Edinburgh ex parte boreali eiusdem dicte diocesis fundatis nobis ex parte eiusdem dicte nostre regine presentatas cum ea qua decuit reuerencia recepisse tenorem qui sequitur continentes.

PIUS episcopus seruus seruorum Dei venerabili fratri episcopo Glasguensi salutem et apostolicam benediccionem. Ad apostolice sedis dignitatis apicem quanquam insufficentibus meritis diuina disposicione vocati ad ea ex suscepte seruitutis officio nostre propensius conuertimus solercie curas per que hospitalium et aliorum piorum locorum necnon pauperum et miserabilium personarum commoditatibus consulatur et ut illa optatum sorciantur effectum cum a nobis petitur fauorem apostolicum libenter impertimur. Sane peticio pro parte carissime in Christo filie nostre Marie Regine Scocie illustris nobis nuper exhibita continebat quod olim postquam felicis recordacionis Nicolaus Papa quintus predecessor noster ex certis tunc expressis causis Hospitale pauperum de Soltre Sanctiandree diocesis in unam dignitatem Cancellariam nuncupatam in ecclesia

and, after its reduction, for the union annexation and incorporation of it with the College and Hospital of the Holy Trinity lately founded near the Burgh of Edinburgh, of the said diocese, on the north side, by the most Serene Princess and Lady, Lady Mary Queen of Scotland, which letters were presented to us on the part of our said Queen, and of which the tenor follows:—

PIUS Bishop, Servant of the servants of God to his venerable brother Bishop of Glasgow, greeting and the apostolic benediction. Called, although with insufficient merits, by divine dispensation to the highest dignity of the Apostolic See, we, on account of the duty of service undertaken by us, the more readily turn our anxious cares to those measures by which provision is made for the advantage of hospitals and other pious places, as well as of poor and miserable persons, and, that they may accomplish the desired end, we willingly impart the apostolic favour when it is sought from us. The petition lately presented to us on behalf of our most dearly beloved daughter in Christ, Mary the illustrious Queen of Scotland, shewed that sometime after our predecessor Pope Nicholas the Fifth of happy memory, for certain causes then expressed, had of apostolic authority by his letters erected the Hospital for the poor at Soltray in the dio-

Sanctiandree, de consensu quondam Alani Caut ipsius hospitalis tunc rectoris, per suas literas apostolica auctoritate erexerat ac de ipsa cancellaria eidem Alano prouiderat, cancellaria predicta que de iure patronatus laicorum Regis Scotorum pro tempore existentis fore noscitur per obitum dicti Alani qui vsque ad vltimum vite sue diem cancellariam ipsam pacifice tenuit ac possedit, et tandem extra Romanam curiam decessit, vacante, clare memorie Jacobus Scotorum Rex dum viueret verus et unicus patronus dicte cancellarie existens in pacifica possessione seu quasi iuris presentandi personam ydoneam ad cancellariam ipsam dum pro tempore vacaret, dilectum filium Johannem Tyry clericum Sanctiandree baccalarium in decretis ad cancellariam ipsam sic vacantem vicario venerabilis fratris nostri episcopi Sanctiandree in spiritualibus generali infra tempus debitum presentauit, ipseque vicarius ad presentacionem huiusmodi prefatum Johannem in cancellariam dicte ecclesie ordinaria auctoritate instituit, ac idem Johannes presentacionis et institucionis huiusmodi vigore dictam cancellariam extitit pacifice assecutus. Cum autem sicut eadem peticio subiungebat prefata Maria Regina ad laudem Omnipotentis

cese of St Andrews into a dignity named the Chancellorship, in the church of St Andrews, with consent of the late Allan Cant then rector of the said hospital, and had provided the said Allan with the said chancellorship; the chancellorship foresaid, which is known to belong by right of laic patronage to the King of Scotland for the time being, becoming vacant by the death of the said Allan, who to the last day of his life peaceably held and possessed the said Chancellorship, and at length died outwith the Roman Court, James King of Scots, of illustrious memory, the true and only patron of the said chancellorship, while he lived being in peaceable possession of the virtual right of presenting a fit person to the said Chancellorship, when it was vacant presented within due time our beloved son John Tyry, clerk of St Andrews, bachelor in canon law, for the said Chancellorship thus vacant, to the Vicar General in things spiritual of our venerable brother the bishop of St Andrews, and the said vicar on that presentation, by his authority as Ordinary inducted the foresaid John in the Chancellorship of the said church, and the said John in virtue of such presentation and induction peaceably holds the said Chancellorship. Since, morover, as the said petition subjoined, the said Queen Mary, of the goods bestowed on her by God, has of new founded and in splendid manner has caused to be constructed and erected for the praise of

Dei vnam Collegiatam Ecclesiam cum Hospitali pauperum prope Burgum de Edinburgh ex parte boriali dicte diocesis pro diuini cultus augmento ac Christi pauperum et aliarum miserabilium personarum recepcione et sustentacione de bonis sibi a Deo collatis de nouo fundauerit, ac egregio quodam opere construi et edificari fecerit, et tam carissimus in Christo filius noster Jacobus modernus Scotorum Rex illustris cuius progenitores ipsum Hospitale de Soltre ad vsum Christi pauperum fundarunt voluntates progenitorum pro posse obseruari facere, quam Regina predicta ut prefatum in cancellariam erectum hospitale ad statum pristinum restituatur ipsique nouo Hospitali erecto perpetuo incorporetur feruenter exoptent, prefatusque Rex ad hoc consensum suum prestare paratus sit, et sicut accepimus ipse Johannes in fauorem vnionis huiusmodi et ut eciam cancellaria ipsa ad hospitale de Soltre sicuti prius erat reducatur, eandem cancellariam quam obtinet sponte et libere resignare proponit, pro parte eiusdem regine asserentis quod fructus redditus et prouentus dicte cancellarie sexaginta librarum sterlingorum secundum communem estimacionem valorem annuum non excedunt nobis fuit humiliter supplicatum ut nomen et dignitatem

Almighty God, a Collegiate Church with an Hospital for the poor near the Burgh of Edinburgh, on the north side, in the said diocese, for the furthering of divine worship, and the reception and maintenance of Christ's poor, and other miserable persons; and both our dearest son in Christ, James the present illustrious King of Scots, whose forefathers founded the said Hospital of Soltray for the use of Christ's poor, in order that the will of his ancestors may as far as possible be observed, and the said Queen, fervently desire that the foresaid Hospital, which has been erected into a Chancellorship, should be restored to its former state, and be incorporated with the said newly erected Hospital for ever, and the foresaid King is prepared to give his consent to this course, and, as we have learned, the said John proposes willingly and freely to resign the said Chancellorship, which he holds, in favour of such union, and also, that the said Chancellorship should revert to the Hospital of Soltray as it formerly was, on the part of the said Queen asserting that according to the common estimation the fruits, rents and proceeds of said chancellorship do not exceed the annual value of sixty pounds sterling, it was humbly supplicated of us, that of our apostolic benignity, we should be pleased to suppress altogether and extinguish the name and dignity of said chancellorship in

huiusmodi cancellarie in prefata ecclesia penitus supprimere et extinguere et hospitale ipsum de Soltre in statum pristinum restituere illudque in omnibus iuribus et pertinenciis prefato nouiter erecto Hospitali perpetuo incorporare vnire et annectere aliasque in premissis oportune prouidere de benignitate apostolica dignaremur. nos qui dudum inter cetera voluimus quod petentes beneficia ecclesiastica aliis vniri teneantur exprimere verum valorem annuum fructuum tam beneficii vniendi quam illius cui vniri petitur alioquin vnio non valeat de premissis tamen certam noticiam non habentes huiusmodi supplicationibus inclinati fraternitati tue per apostolica scripta mandamus quatenus vocatis episcopo predicto necnon dilectis filiis capitulo dicte ecclesie Sanctiandree et aliis quorum interest de premissis omnibus et singulis eorumque circumstanciis vniuersis auctoritate nostra te diligenter informes et si per informacionem huiusmodi ita esse repereris ab ipso Johanne vel procuratore suo ad hoc ab eo specialiter constituto resignacionem huiusmodi si illam in tuis manibus sponte et libere facere voluerit ut profertur auctoritate nostra hac vice duntaxat recipias et

said church, and restore the said Hospital of Soltray to its pristine state, and to incorporate, unite and annex it in all its rights and pertinents to the foresaid newly erected Hospital for ever, and otherwise to make suitable provisions in the premises. We, who long ago have among other things resolved, that those petitioning for ecclesiastical benefices to be united to others should be bound to declare the true annual value of the fruits, as well of the benefice to be united, as of that to which it is sought to be united, otherwise the union to be of none effect, although not having certain notice of the premises, being favourably disposed to these supplications, do by our apostolic writings, charge your fraternity to call the Bishop aforesaid and also our beloved sons the chapter of said church of St Andrews and others interested, and on our authority diligently to inform yourself of all and sundry the premises, and of their whole circumstances, and, if by such information you find it so to be, to receive and accept from the said John or his procurator specially constituted by him for this purpose, such resignation, if he shall voluntarily and freely determine to make the same in your hands as aforesaid, by our authority granted for this occasion only, and the said resignation being received and accepted by you, to

admittas eaque per te recepta et admissa cancellariam ipsam cum per huiusmodi resignacionem vacauerit dummodo tempore date presencium non sit in ea alias alicui specialiter ius quesitum ac prefati Jacobi moderni seu pro tempore existentis Regis Scotorum ad id expressus accedat assensus nomine et dignitate ipsius cancellarie in dicta ecclesia per te suppressis penitus et extinctis eodemque Hospitali de Soltre in statum pristinum reducto illud cum omnibus iuribus et pertinenciis suis nouo Hospitali huiusmodi cuius et eiusdem ecclesie fructuum reddituum et prouentuum veros annuos valores presentibus haberi volumus pro expressis eadem auctoritate nostra perpetuo vnias annectas et incorpores ita quod liceat ex tunc rectori dicti noui Hospitalis per se vel alium seu alios corporaliter Hospitalis de Soltre iuriumque et pertinenciarum predictorum possessionem propria auctoritate apprehendere et perpetuo retinere ac illius fructus redditus et prouentus in sustentacionem pauperum et infirmorum ac alias in vtilitatem ipsorum hospitalium iuxta formam constitucionis pie memorie Clementis pape V. eciam predecessoris nostri super hoc in consilio Viennensi edite com-

take over the said Chancellorship, when by such resignation it shall be vacant, provided at the date of these presents no one shall have otherwise acquired a special right in it, and if the express consent to that effect of the foresaid James, who now is, or of the King of Scots for the time being, shall be accorded, and, having totally suppressed and extinguished the name and dignity of the said Chancellorship in the said church, and having reduced the said Hospital of Soltray to its pristine state, by our said authority, to unite, annex, and incorporate for ever the same with all its rights and pertinents to the said new Hospital, the true annual values of the fruits, rents and profits of that Hospital and of that church, we will shall be held for expressed in these presents, so that it may be lawful thenceforth for the Rector of the said new Hospital by himself or others one or more on his own authority to take and retain for ever corporal possession of the Hospital of Soltray and its rights and pertinents aforesaid, and to receive its fruits, rents and proceeds, to be applied for the maintenance of the poor and the infirm, and otherwise for the use of said hospitals according to the form of constitution of Pope Clement V., of pious memory, our predecessor, given forth regarding this matter in the council

mittendos percipere diocesani loci et cuiusuis alterius licencia super hoc minime requisita non obstantibus constitucionibus et ordinacionibus apostolicis ac predicta nostra voluntate necnon statutis et consuetudinibus eiusdem ecclesie iuramento confirmacione apostolica vel quauis firmitate alia roboratis contrariis quibuscunque aut si aliquis super prouisionibus sibi faciendis et dignitatibus ipsius ecclesie speciales vel aliis beneficiis ecclesiasticis in illis partibus generales dicte sedis vel legatorum eius literas impetrauerit eciam per eas ad inhibitionem resignacionem et decretum vel alias quomodolibet sit processum quas quidem literas et processus habitos per easdem ac quecunque inde secuta ad dictam cancellariam volumus non extendi sed nullum per hoc eis quoad assecucionem dignitatum vel beneficiorum aliorum preiudicium generari et quibuslibet priuilegiis indulgenciis et literis apostolicis generalibus vel specialibus quorumcunque tenorum existant per que presentibus non expressa vel totaliter non inserta effectus earum impediri valeat quomodolibet vel differri et de quibus quorumcunque totis tenoribus de verbo ad verbum habenda sit in nostris literis mencio

of Vienne, the permission of the diocesan of the place or of any other whatever to this effect not being in any way required, notwithstanding constitutions and ordinances apostolic, and our foresaid will as also the statutes and customs of the said church strengthened by oath, confirmation apostolic, or any other security to the contrary whatsoever, or though any one shall have obtained special letters for the provisions to be made to himself and dignities of the said church, or general letters of the said See or its Legates to other ecclesiastical benefices in those parts, even though in virtue of these, or in any other way whatsoever, the matter has gone the length of inhibition, resignation and decree; which letters indeed and proceedings taken in virtue of the same and all that follows thereupon, we ordain not to be extended to the said chancellorship, but that no prejudice be caused to them by this provision as to the acquisition of other dignities or benefices, and notwithstanding whatsoever privileges indulgences and apostolic letters, general or special of whatsoever tenor they be, not expressed or altogether not inserted in the present letters, by which their effect can be impeded in any manner or delayed, and of which, and the whole tenors of which, special mention should be made word for word

specialis prouiso quod propter vnionem anneccionem et incorporacionem huiusmodi si vigore presencium fiant et effectum sorciantur predictum hospitale de Soltre debitis non fraudetur obsequiis sed illius congrue supportentur onera consueta. Ceterum cum non sit verisimile quod quis beneficia sua multis forsan laboribus acquisita ex quibus percipit vite subsidium absque magna causa sponte resignet attente prouideas ne in resignacione huiusmodi si fiat aliqua prauitas interueniat seu eciam corruptela nos enim si extinctionem suppressionem reduccionem vnionem anneccionem et incorporacionem huiusmodi per te vigore presencium fieri contigerit ex nunc irritum decernimus et inane si secus super hiis a quoquam quauis auctoritate scienter vel ignoranter contigerit attemptari. Datum Rome apud Sanctum Petrum anno incarnacionis Dominice millesimo quadringentesimo sexagesimo decimo kalendas Nouembris pontificatus nostri anno tercio.

Nos vero Andreas episcopus iudex et executor prefatus tanquam obediens filius volentes mandatum apostolicum nobis in hac parte directum reuer-

in our letters, provided that on account of such union, annexation, and incorporation, if they shall, in virtue of these presents, be brought about and receive effect, the foresaid Hospital of Soltray shall not be defrauded of its due services, but its accustomed burdens shall be suitably supported. But since it is not likely, that any one will, without good cause, of his own accord will resign his benefices acquired perhaps with much labour and from which he obtains the support of his life, you shall make careful provision, that in such resignation, if it shall be made, no wrong or corruption shall happen; for if it come about that such extinction, suppression, reduction, union, annexation, and incorporation be made by you in virtue of these presents, we declare it henceforth null and void if any attempt towards this has been wrongfully made by any one, by any authority whatever, either knowingly or ignorantly. Given at Rome at St Peter's in the year of the incarnation of our Lord one thousand four hundred and sixty, the tenth day before the calends of November, and of our Pontificate the third year.

And we Andrew, Bishop, judge and executor aforesaid, as an obedient son wishing reverently to carry out the apostolic commission, directed to us in this

enter exequi ut tenemur ad prefate serenissime principisse domine nostre regine suorumque procuratorum hoc requirencium instanciam reuerendo in Christo patri et domino domino Jacobo miseracione diuina episcopo Sanctiandree venerabilibusque patre et viris priore et conuentu Sanctiandree ac aliis interesse habentibus per nos debite citatis atque vocatis ipsorum claro et expresso consensu ad infrascripta accedente diligente informacione omnium singulorumque in prefatis literis apostolicis contentorum et relatorum per nos habita terminis substancialibus in causa et materia prefatis de iure seruandis debite seruatis conclusioneque in eadem facta ad nostre sentencie diffinitiue in causa et materia prefatis prolacionem processimus eamque tulimus et promulgauimus forma subsequente. Nos Andreas Dei et apostolice sedis gracia episcopus Glasguensis iudex et executor vnicus a sede apostolica cause et partibus infrascriptis specialiter deputati pro tribunali sedentes in quadam causa super extinccione et suppressione nominis et dignitatis cancellarie Sanctiandree alias erectorum et reduccione Hospitalis pauperum de Soltre dicte diocesis in statum pristinum illoque reducto

part, as we are bound to do, at the instance of the said most serene Princess our Lady the Queen and her procurators so demanding, the reverend father and lord in Christ Lord James by divine mercy Bishop of St Andrews, and the venerable father and men the Prior and the Convent of St Andrews, and others having interest having been duly cited and called by us, and their clear and express consent to the underwritten being accorded, diligent information of all and sundry in the foresaid letters apostolic contained and related being obtained by us, the substantial limits which ought of right to be observed in the cause and matter foresaid having been duly observed, and a conclusion in the same having been formed, we have proceeded to pronounce our definite sentence in the cause and matter aforesaid, and have passed and promulgated the same in the following form: We Andrew, by the grace of God and the Apostolic See Bishop of Glasgow, judge and sole executor to the cause and to the parties underwritten specially deputed by the Apostolic See, sitting in judgment in a certain cause regarding the extinction and suppression of the name and dignity of the Chancellorship of St Andrews formerly established, and of the reduction of the Hospital of the poor of Soltray of said diocese to its pristine state, and after its reduc-

vnione anneccione et incorporacione Collegio et Hospitali nouiter per serenissimam principissam et dominam dominam Mariam Dei gracia Reginam Scocie illustrissimam iuxta Burgum de Edinburgh ex parte boriali eiusdem fundato per dictam serenissimam principissam actricem ab vna et reuerendum in Christo patrem et dominum Jacobum miseracione diuina episcopum Sanctiandree venerabilesque patrem et viros priorem et conuentum eiusdem et alios quorum interest reos partibus ab altera coram nobis iudicialiter ventilata cognoscentes auditis prius pro parte dicte domine Regine propositis iuribus et probacionibus ipsius ex parte productis visis rimatis et intellectis ac ad plenum discussis et consideratis et quia omnia et singula in huiusmodi literis apostolicis narrata et expressa veritate fulciri comperimus ac resignacionem prefate cancellarie per dictum Johannem possessorem eiusdem alias sponte et libere factam consensuque serenissimi principis et domini nostri domini Jacobi moderni Scotorum Regis illustrissimi dicti hospitalis de Soltre patroni vnici et indubitati ad hoc reuerendi eciam in Christo patris et domini domini Jacobi episcopi Sanctiandree prioris

tion, for its union, annexation, and incorporation with the College and Hospital newly founded by the most serene Princess and lady Lady Mary by the grace of God the most illustrious Queen of Scotland, near the Burgh of Edinburgh on the north side of the same, examining the pleas raised judicially before us by the said Most Serene Princess, pursuer, on the one part, and by the reverend father in Christ and Lord James by divine mercy Bishop of St Andrews, and the venerable father and men the Prior and the Convent of the same, and others interested, defenders, on the other part, having first heard what was set forth on the part of the said Lady the Queen, and having seen, investigated, and understood and fully discussed and considered the rights and proofs brought forward on her part, and whereas we have found the matters all and sundry narrated and expressed in the said Letters Apostolic to be founded in truth, and the resignation of the foresaid Chancellorship to have been freely and spontaneously made by the said John, the possessor of the same, and with the consent of the most Serene Prince and our lord Lord James the present most illustrious King of Scots, sole and undoubted patron of the said hospital of Soltray, as also with the clear express consent of the reverend father in Christ and lord, Lord James

et conuentus euisdem prefatorum claro et expresso accedente consensu Deum pre oculis habentes eiusque nomine sanctissimo primitus inuocato iurisperitorum eciam consilio secuto auctoritate apostolica qua fungimur in hac parte per hanc nostram sentenciam diffinitiuam quam ferimus in hiis scriptis pronunciamus decernimus et declaramus nomen et dignitatem dicte cancellarie in dicta ecclesia Sanctiandree extinguenda et suppremenda fore prout supprimimus et extinguimus et Hospitale pauperum de Soltre in statum pristinum reducendum fore et reducimus et sic reductum cum omnibus iuribus et pertinenciis suis Collegio et Hospitali Sancte Trinitatis nouiter fundato perpetuo vnimus annectimus et incorporamus aliasque in premissis et eorum circumstanciis prouidendum fore decernimus. Lecta lata et promulgata fuit hec nostra sentencia diffinitiua in ecclesia parochiali Beati Michaelis de Linlithqw Sanctiandree diocesis sexto die mensis Marcii anno Domini millesimo quadringentesimo sexagesimo primo indiccione decima pontificatus sanctissimi in Christo patris et domini nostri domini Pii diuina prouidencia Pape secundi predicti anno quarto. In quorum omnium et singulorum fidem et testimonium premissorum has literas nostras

Bishop of St Andrews, the Prior and the Convent of the same aforesaid, having God before our eyes and having first invoked his most holy name, having followed also the advice of those skilled in law, by the apostolic authority which we exercise in this part, by this our definite sentence which we pass in these writings, pronounce, decern, and declare the name and dignity of the said Chancellorship in the said Church of St Andrews to be extinguished and suppressed, as we suppress and extinguish the same, and the Hospital of the poor of Soltray to be reduced to its pristine state, and we reduce it; and it being so reduced we unite, annex, and incorporate it for ever with all its rights and pertinents, with the newly founded College and Hospital of the Holy Trinity, and otherwise decern provision to be made in the premises and their circumstances: This our definite sentence was read, passed, and promulgated in the Parish Church of St Michael of Linlithgow, in the diocese of St Andrews, on the sixth day of the month of March, in the year of our Lord one thousand four hundred and sixty-one, the tenth indiction, of the pontificate of the most holy father in Christ and our Lord, by divine providence Pope Pius the Second aforesaid, the

sigillo nostro roboratas signo et subscripcione notarii publici subscripti subscribi et publicari mandauimus. Datum et actum apud burgum de Lynlithqw predictum anno die mense indiccione et pontificatu prefatis presentibus ibidem discretis et honestis viris dominis Thoma de Kirkpatrik, Patricio Broune, Johanne Pettygrew, presbyteris Jacobo Stewart, Johanne Neilsone, Andrea Grynlaw et Georgio Clerc, testibus ad premissa vocatis et rogatis.

Et ego Ricardus Roberti presbyter Sanctiandree diocesis publicus auctoritate imperiali notarius et coram dicto reuerendo patre domino episcopo iudice in huiusmodi causa scriba iuratus quia dictarum literarum apostolicarum presentacionem recepcionem causeque huiusmodi deduccionem et sentencie prolacionem ceterisque omnibus et singulis dum sic ut premittitur dicerentur et fierent vnacum prenominatis testibus presens interfui eaque omnia et singula sic fieri et dici vidi et audiui ac in notam

fourth year. In faith and testimony of all and each of these premises we have ordered these our letters, confirmed by our seal, to be subscribed with the seal and subscription of the notary underwritten and to be published. Given and done at the burgh of Linlithgow aforesaid, the year, day, month, indiction, and pontificate aforesaid, in presence of discreet and honest men Sirs Thomas of Kirkpatrick, Patrick Brown, John Pettigrew, Priests, James Stewart, John Neilson, Andrew Greenlaw, and George Clark, witnesses to the premises called and required.

And I Richard Roberts, priest of the diocese of St Andrews, by imperial authority, Notary Public, and in presence of the said reverend father, the Lord Bishop, judge in this cause, sworn scribe, inasmuch as at the presentation and reception of the said letters apostolic, the deduction of said cause, and the passing of sentence, and the other things all and sundry while as is premised, they were said and done, along with the forenamed witnesses, being present, I saw and heard them all and each so done and spoken, and took a note of the same, and have then made out this

recepi ideoque hoc presens publicum instrumentum siue presentes literas huiusmodi processum in se continentes siue continens exinde confeci et in hanc publicam formam redegi signoque et nomine meis solitis et consuetis vnacum appensione sigilli dicti domini episcopi et iudicis signaui rogatus et requisitus in fidem et testimonium omnium et singulorum premissorum.

present public instrument or the present letters containing in themselves such process, and have reduced them into this public form, and have signed them with my name and seal, used and wont, as well as affixed the seal of the said Lord Bishop and Judge, being called and required in faith and testimony of all and sundry the premises.

II.

CONFIRMATION by James Kennedy, Archbishop of St Andrews, of the Foundation of the Collegiate Church and Hospital of the Holy Trinity, St Andrews, 1st April 1462.

JACOBUS Dei et apostolice sedis gracia Sanctiandree episcopus vniuersis sancte matris ecclesie filiis ad quorum noticias presentes litere peruenerint salutem in omnium Saluatore. Splendor eterne

JAMES, by the grace of God and the Apostolic See, Bishop of St Andrews; To all the sons of holy mother church to whose notice the present letters shall come, greeting in the Saviour of all. The splendour of the eternal glory which with

glorie qui sua mundum illuminat ineffabili claritate pia vota fidelium de sue maiestatis clemencia tum benigno fauore prosequitur cum deuota ipsorum humilitas et sincerus affectus in diuini cultus augmentum feruore dinoscuntur. Nos vero de superis ad inferiora deriuatis exemplis piis desideriis iustisque supplicancium precibus quos fides spes et caritas ad deuocionis affectum continue solicitant ut ad graciam dispositi et ad gloriam inuitati in piis actibus accuracius perseuerent tam vigor et feruor equitatis quam ordo racionis fauere nos inducunt. Literas igitur prepotentissime principisse et domine domine Marie Die gracia Regine Scocie suo sigillo sigillatas pium et laudabile eiusdem principisse Propositum perpetuum Collegium siue Ecclesiam Collegiatam vnius prepositi et octo capellanorum ac duorum puerorum ibidem diuino cultui deputandorum prope Burgum de Edinburgh ex parte boreali eiusdem cum Hospitali tresdecim pauperum in eodem sustinendorum perpetuis futuris temporibus duraturum super terris et ecclesiis beneficii de Soltre Collegio et Hospitali suo apostolica auctoritate vnitis et incorporatis necnon et super terris de Ballernow et hospitali de

its ineffable brightness illuminates the world, of the clemency of its majesty accompanies with benignant favour the pious vows of the faithful when their devout humility and sincere zeal for the furthering of divine worship are distinguished for fervour. We, therefore, deriving example for things below from those above, are induced, both by the vigour and zeal of equity and the course of reason, to favour the pious desires and earnest prayers of supplicants whom faith, hope, and charity continually stir up to zeal of devotion, so that being disposed to grace and invited to glory they may the more diligently persevere in pious actions. We have received the letters of that most potent Princess and lady, Lady Mary, by the grace of God Queen of Scotland, sealed with her seal, containing the pious and laudable proposal of the said princess, desiring to constitute and found a perpetual College or Collegiate Church, of one provost and eight chaplains and two boys to perform divine worship therein, near the Burgh of Edinburgh, on the north side of the same, with an Hospital for the maintenance therein in all time coming of thirteen poor persons, from the lands and churches of the benefice of Soltray by apostolic authority united to and incorporated with the said College and its Hospital, and also from the lands of Ballerno and the hospital of Uthirrogale, and its other lands,

Vthirregale ac aliis terris suis possessionibus et annuis redditibus infrascriptis ordinare et fundare desiderantis necnon ipsius principisse humilem supplicacionem vt dicti Collegii fundacionem disposicionem et ordinacionem prefatas per nostri solicitudinem officii ad effectum producere approbare ratificare et confirmare dignaremur continentes recepimus in hec verba.

MARIA Dei gracia Regina Scocie reuerendo in Christo patri ac domino domino Jacobo Dei et apoctolice sedis gracia Sanctiandree episcopo etc. consanguineo nostro carissimo reuerencias tanto patre dignas cum honore. Nouerit vestra reuerenda paternitas quod in laudem et honorem Sancte Trinitatis beate et gloriose semper Virginis Marie Sancte Niniani confessoris et omnium sanctorum et electorum Dei nos Maria predicta cum consensu et assensu illustrissimi principis Domini Jacobi inuictissimi filii nostri Regis Scocie ad perpetuam rei memoriam pro salute anime quondam inclitissimi principis Jacobi Scotorum Regis pie memorie nostri sponsi necnon omnium regum et reginarum defunctorum et defunctarum Scocie necnon pro salute illustrissimi principis Jacobi moderni Regis Scocie filii nostri prefati et pro salute anime nostre patris et matris omnium nostrorum

possessions and annual rents underwritten, likewise the humble supplication of the said Princess, that of the solicitude of our office we should see fit to carry into effect, approve, ratify, and confirm the foresaid foundation, disposition and settlement of the said College, in these words:

MARY, by the grace of God Queen of Scotland, to the reverend father in Christ and lord, Lord James, by the grace of God and the Apostolic See Bishop of St Andrews, &c., to our most dearly beloved cousin, the reverence and honour worthy of such a father: Know reverend father that, for the praise and honour of the Holy Trinity, of the blessed and ever glorious virgin Mary, of St Ninian the confessor, and of all the saints and elect of God, we Mary aforesaid, with consent and assent of the most illustrious prince Lord James, our most invincible son, King of Scotland, in perpetual memorial hereof, for the weal of the soul of the late most famous prince, James King of Scots, of pious memory, our husband, as well as of all the Kings and Queens of Scotland deceased, and for the health of the most illustrious prince, James the present King of Scotland, our son aforesaid, and for the weal of our soul and of the souls of our father and mother, of all our

antecessorum filiorum et filiarum successorum et ab eisdem descendencium ac pro salute reuerendi in Christo patris Domini Jacobi Sanctiandree episcopi moderni consanguinei nostri carissimi et omnium consanguinitate affinitate seu beneficiis nobis coniunctorum et omnium quos in hac vita offendimus ad quorum obligamur emendam ac omnium fidelium defunctorum PREPOSITURAM pro preposito qui aliis in preeminencia honore et dignitate in Ecclessia Collegiata Sancte Trinitatis prope Edinburgh quoad chori et diuini cultus regimen preponetur et octo prebendis perpetuis in quibus octo presbyteri Deo perpetuo administraturi cum duobus pueris siue clericis cum sufficiente porcione inferius designanda deputabuntur facimus constituimus et ordinamus ac in perpetuum fundamus. Preposituram vero dicte ecclesie Collegiate Sancte Trinitatis prope Edinburgh super fructibus inferius designatis cum modificacione sequenti deputamus et assignamus. Prepositus vero dicti Collegii habebit pro sua sustentacione ecclesiam de Soltre et subibit onera dicte ecclesie incumbencia videlicet soluet pensionem vicario dicte ecclesie debitam et sustentabit tres pauperes ibidem degentes cum reparacione et sustentacione ecclesie in tecto et ornamentis ut oportet .

ancestors, of the sons and daughters, the successors and those descending from them, and for the health of the reverend father in Christ, Lord James, present Bishop of St Andrews, our dearest cousin, and of all connected with us by consanguinity, affinity, or benefits, and of all whom we have offended in this life, to whom we are bound to make amends, and of all the faithful deceased, make, constitute, and ordain and for ever found a PROVOSTRY for a Provost, who shall be set over the others in pre-eminence, honour, and dignity in the Collegiate Church of the Holy Trinity near Edinburgh, as regards the regulation of the choir and of divine worship, with eight perpetual Prebends, in which eight Priests to serve God for ever, with two boys or clerks, with a sufficient maintenance to be specified below, shall be appointed. We moreover destine and assign the fruits below specified, under the following modification, to the Provostry of the said Collegiate Church of the Holy Trinity near Edinburgh: The Provost of the said college shall have for his maintenance the church of Soltray, and shall undertake the burdens incumbent on the said church, viz., he shall pay the pension due to the Vicar of the said church, and shall sustain three poor persons living there, along with the repair and upholding of the church in its roof and ornaments as is fitting. He shall also have the lands of

qui eciam habebit terras de lie Barnis de Soltre cum pertinenciis suis et terras ville de Hangandschaw cum suis pertinenciis dari solitis et consuetis et ecclesiam de Lempetlaw cum vniuersis suis fructibus eidem pertinentibus . et pro istis duabus ecclesiis preposito assignatis ipse prepositus subibit episcopalia et archidiaconalia onera de huiusmodi ecclesiis prestari solita et consueta. Primus vero prebendarius post prepositum vocabitur magister hospitalis Sancte Trinitatis prope Edinburgh qui habebit pro sua prebenda et sustentacione congrua quartam partem fructuum rectorie ecclesie de Strathichmartin Sanctiandree diocesis et duas libratas terrarum in villa de Fawlahill infra Heriotmoore et duas marcas annui redditus de domo quondam Willelmi Clunes in Leythe et viginti solidos in villa de Risolton et quinque solidos annui redditus de domibus Joannis Alansone et Joannis Lawsone in Leyth ad duos anni terminos Sancti Martini et Penthecostes per equales porciones et viginti solidos annui redditus de domo Wauklet in Edinburgo et quinque solidos de domo domini Thome episcopi Dunkeldensis et sex solidos et octo denarios in villa de Lawdre per nos postea limitandos et sex solidos octo denarios in villa de Stramiglow per nos postea de-

the Barns of Soltray with their pertinents, and the lands of the town of Hangandshaw with their pertinents used and wont to be given, and the church of Lempetlaw with all the fruits thereto belonging; and for these two churches assigned to the said Provost, he shall bear the Episcopal and Archidiaconal dues of the said churches used and wont to be paid. The first prebendary after the Provost, shall be called Master of the Hospital of the Holy Trinity near Edinburgh, and shall have for his prebend and suitable maintenance a fourth part of the fruits of the rectorial church of Strathmartin, in the diocese of St Andrews, and two pounds land in the town of Fawlahill within Heriot moor, and two merks of annual rent from the house of the late William Clunes in Leith, and twenty shillings in the town of Risolton, and five shillings of annual rent from the houses of John Alanson and John Lawson in Leith, at two terms in the year Martinmas and Whitsunday by equal portions, and twenty shillings of annual rent from the house of Wauklet in Edinburgh, and five shillings from the house of the Lord Thomas, Bishop of Dunkeld, and six shillings and eight pence in the town of Lawder to be afterwards bounded by us, and six shillings and eight pence in the town of Strathmiglow to be by us afterwards declared, as in the rental of

clarandos prout in rentali de Soltre clarius continetur et decem solidos de quadam villa prope Linlithcow prout in rentali continetur et quinque marcatas terrarum de Browderstanis et de Gilestoun infra dominium de Soltre per nos cum suis limitibus terminis et bondis specificandis . qui Magister Hospitalis habebit plenam disposicionem omnium fructuum hospitali et pauperibus inferius designandorum pro eorundem sustentacione limitandorum sic quod prouide et prudentur omnibus et vnicuique in singulis necessariis secundum Deum et bonam consciensiam et fructuum facultates prouideat bis in anno preposito et capitulo ad duos anni terminos Penthecostes et Sancti Martini in hieme et aliis temporibus si necessitatas id exigerit computum prestando et reddendo. Secundus prebendarius post prepositum [qui] vocabitur Sacrista habebit pro sua sustentacione quinque marcatas terrarum in villa de Hill infra dominium de Ballernow et quinque marcatas terrarum de Browderstanis [et] de Gilston dominii de Soltre per nos limitandas vt supra et quartam partem fructuum rectorie de Strathichinmartin . qui Sacrista habebit plenam disposicionem omnium fructuum inferius designandorum communitati et capitulo pertinencium pro necessariis cotidianis ecclesie

Soltray is more clearly contained, and ten shillings from a certain town near Linlithgow, as is contained in the rental, and five merks of the lands of Browderstanes and Gileston, within the lordship of Soltray, with their limits, marches, and bounds to be by us specified; and the said Master of the Hospital shall have the full disposal of all the fruits to be described below for the Hospital and the poor, and to be appointed for their maintenance, so that he may providently and prudently make provision for all and each in all necessaries according to God and a good conscience, and the capabilities of the fruits, shewing and rendering an account to the provost and chapter twice a-year, at the two terms of Whitsunday and Martinmas in winter and at other times, if necessity shall so require. The second Prebendary after the Provost, who shall be called the Sacristan, shall have for his maintenance five merks of lands in the town of Hill, within the lordship of Ballerno, and five merks of the lands of Browderstanes and Gileston, of the lordship of Soltray, to be defined by us as above, and a fourth part of the profits of the rectory of Strathmartin: And the said Sacristan shall have the full disposal of all the fruits to be below described pertaining to the Community and Chapter, to be collected and disposed of for the daily necessaries of the collegiate church;

collegiate comparandorum et disponendorum . qui reddet computum preposito et prebendariis quater in anno in quatuor sabbatis quatuor temporum et non comparabit preciosa negocia nisi cum consilio prepositi et capituli custodietque ecclesiam in omni honestate ornamenta eiusdem iocalia et vasa sacra seruabit campanas pulsabit vinum panem et luminaria ministrabit aliaque exercebit que de consuetudine laudabili et obseruata in aliis ecclesiis ad officium Sacriste pertinere dinoscuntur. Tercius Prebendarius qui vocabitur Prebendarius de Browderstanis habebit pro sua prebenda quinque marcatas terrarum de Browderstanis et Gilstoun limitandas per nos et aliam quartam partem fructuum rectorie de Strathichinmartin antedicte. Quartus prebendarius qui intitulabitur Prebendarius de Strachinmartin habebit quinque marcatas terrarum de Browderstanis et de Gilstoun per nos limitandas et vltimam quartam [partem] fructuum rectorie de Strathichinmartin antedicte. Qui quatuor prebendarii prenominati subibunt omnia onera episcopalia et archidiaconalia et alia de iure solui consueta cum ecclesie reparacione et sustentacione de primis fructibus rectorie de Strathichinmartin inter predictos quatuor prebendarios equaliter con-

and he shall render an account to the Provost and Prebendaries four times in the year, on the four Saturdays of the four seasons, and shall not settle important business unless with the advice of the Provost and Chapter, and shall keep the church in all propriety, and have the custody of the ornaments, jewels, and sacred vessels of the same, and ring the bells, and supply wine, bread and lights, and exercise the other functions which, according to laudable and observed usage in other churches, are known to belong to the office of Sacristan. The third Prebendary, who shall be called Prebendary of Browderstanes, shall have for his prebend five merks of the lands of Browderstanes and Gilston, to be defined by us, and another fourth part of the fruits of the rectory of Strathmartin aforesaid. The fourth Prebendary, who shall be styled Prebendary of Strathmartin, shall have five merks of the lands of Browderstanes and of Gilston, to be defined by us, and the last fourth part of the fruits of the rectory of Strathmartin aforesaid: And the four Prebendaries aforenamed shall undertake all the dues which of right used to be paid to the Bishop, Archdeacon, and others, with the repair and upholding of the church from the first fruits of the rectory of Strathmartin equally divided among the foresaid four Prebendaries. The fifth Prebendary,

dinisis. Quintus prebendarius qui intitulabitur de Gilstoun habebit quinque marcatas terrarum de Browderstanis et Gilstoun per nos limitandas vt supra et primam quartam [partem] fructuum rectorie ecclesie de Ormistoun Sanctiandree diocesis. Sextus prebendarius qui intitulabitur de Ormistoun habebit quinque marcatas terrarum de Browderstanis et Gilstoun per nos limitandas et secundam quartam partem fructuum rectorie ecclesie de Ormistoun antedicte . Septimus prebendarius qui intitulabitur de Hill habebit quinque marcatas terrarum infra dominium de Ballerno per nos limitandas et terciam quartam [partem] fructuum rectorie de Ormistoun. Octauus prebendarius qui intitulabitur de Newlandis habebit quinque marcatas terrarum de Newlandis dominii de Soltre et vltimam quartam [partem] fructuum rectorie de Ormistoun antedicte. Quiquidem vltimi quatuor prebendarii subibunt onera episcopalia et archidiaconalia cum reparacione et sustentacione ecclesie de Ormistoun de primis fructibus eiusdem cum aliis oneribus de iure et consuetudine solui consuetis. Statuimus insuper et ordinamus vt in eodem nostro Collegio sint duo Clerici qui parebunt preceptis prepositi et habebunt pro sua sustentacione decem libratas terrarum de Ballerno

who shall be styled of Gilston, shall have five merks of the lands of Browderstanes and Gilston, to be defined by us as above, and the first fourth part of the fruits of the rectory of the church of Ormiston, in the diocese of St Andrews. The sixth Prebendary, who shall be styled of Ormiston, shall have five merks of the lands of Browderstanes and Gilston, to be defined by us, and the second fourth part of the fruits of the rectory of the church of Ormiston aforesaid. The seventh Prebendary, who shall be styled of Hill, shall have five merks of lands within the lordship of Ballerno, to be defined by us, and the third fourth part of the fruits of the rectory of Ormiston. The eighth Prebendary, who shall be styled of Newlands, shall have five merks of the lands of Newlands, in the lordship of Soultray, and the last fourth part of the fruits of the rectory of Ormiston aforesaid. The last four Prebendaries shall bear the dues to the Bishop and Archdeacon, with the repair and upholding of the church of Ormiston from the first fruits of the same, with the other burdens wont to be discharged of law and custom. We moreover statute and ordain, that in our said College there shall be two clerks, who shall obey the commands of the Provost, and shall have for their maintenance ten pounds of the lands of Ballerno, to be defined by us, and to be equally

per nos limitandas equaliter inter eosdem diuidendas qui remouebiles erunt ad placitum prepositi et collegii antedictorum. Et pro tresdecim pauperibus in nostro Hospitali sustinendis limitamus et ordinamus hospitale de Vthirrogill et rectoriam de Weymis Sanctiandree diocesis et decem libras annui redditus ville de Edinburgh de communi eiusdem ac decem libras de certis terris et tenementis cum acris et annuis redditibus nobis in villa de Leyth debitis et pertinentibus . Insuper ordinamus pro reparacione Ecclesie Collegiate prefate et necessariis in eadem supportandis quadraginta sex libras nouem solidos de fructibus rectorie de Kirkurd Glasguensis diocesis [et] resta terrarum nostrarum de Ballerno. Onera vero episcopalia archidiaconalia et alia solui consueta cum reparacione et sustentacione ecclesie de Kirkurd subibit Sacrista dicti collegii qui habebit recepcionem omnium bonorum prescriptorum vt superius est expressum. Quiquidem prebendarii predicti omni die anni matutinas magnam missam vesperas [et] completorium cum nota cantabunt. Prepositus vero matutinis misse [et] vesperis festiuis diebus interesse teneatur. Omnes autem prebendarii predicti residenciam facient personalem et per seipsos et non per alium seu alios premissa

divided between them, and who shall be removeable at the pleasure of the Provost and College aforesaid. And for the maintenance of thirteen poor persons in our hospital we set aside and ordain the Hospital of Uthirrogill and the rectory of Wemyss, in the diocese of St Andrews, and ten pounds of the annual rents of the town of Edinburgh from the common good of the same, and ten pounds from certain lands and tenements, with acres and annual rents due and pertaining to us in the town of Leith. Moreover, we ordain for the repair of the Collegiate Church aforesaid, and for the supply of necessaries in the same, forty-six pounds nine shillings of the profits of the rectory of Kirkurd, in the diocese of Glasgow, and the remainder of our lands of Ballerno; and the Sacristan of the said Collegiate Church shall bear the burdens of the dues used to be paid to the Bishop, Archdeacon and others, with the repair and upholding of said church of Kirkurd, and shall have the receiving of all the goods aforesaid as is above expressed. The Prebendaries aforesaid, moreover, shall every day in the year sing matins, high mass, vespers, and compline, with notes. The Provost shall also be present at matins, mass, and vespers on festival days. All the foresaid prebendaries shall also make personal residence, and shall implement and discharge the duties aforesaid by themselves

perimplebunt siue administrabunt. Ipse vero prepositus ad residenciam obligetur personalem ita quod si ipsum per quindecim dies continuos abesse contingat capitulum dicti Collegii vt ad residenciam reuertatur apud patronum instabit vt ipsum ad residenciam compellat et si infra quindecim dies alios post denunciacionem patrono factam ad residenciam non reuertatur animo permanenti ex hoc ipso ipsa prepositura sine strepitu et figura iudicis vacare censebitur et si infra quindecim dies post huiusmodi vacacionem patronus ordinario prepositum nouum non presentauerit capitulum alium prepositum ordinario presentabit. Et ordinamus quod nullus prebendariorum aut clericorum sine licencia prepositi desuper obtenta se absentet. vltra vero quindecim dies preposito aliquem licenciare non licebit nisi ex magna et ardua causa et hoc cum consensu capituli sui desuper requisito et obtento. contrarium autem faciens ipso facto a iure suo prebende cadet et sua prebenda censebitur vacare et realiter vacet per prepositum et capitulum et prouisionem ordinarii libere conferenda. Et si quis prebendariorum concubinam seu focariam detinuerit et illam per prepositum ter monitus non dimiserit cum effectu eo viso censeatur sua prebenda vacare et

personally, and not by others one or more. The Provost himself, shall likewise be bound to personal residence, in such manner that if he shall happen to be absent for fifteen successive days, the Chapter of the said College, shall, in order that he may return to personal residence, urge upon the Patron to compel him to residence, and if within fifteen other days after denunciation made to the patron, he shall not return to residence with the intention of remaining, the said provostship shall, from that fact alone, be considered vacant without trouble or formal deliverance of the judge; and if within fifteen days after such vacancy the Patron shall not have presented a new Provost to the Ordinary, the Chapter shall present another Provost to the Ordinary. And we ordain that none of the prebendaries or clerks shall absent himself without leave of the Provost first obtained; but it shall not be lawful for the Provost to licence any one beyond fifteen days, unless for great and pressing reason, and this with consent of the Chapter thereto required and obtained. One doing the contrary shall *ipso facto* fall from his right of prebend, and his prebend shall be considered, and shall be in reality vacant, and at the free disposal of the Provost and chapter, and the provision of the Ordinary. And if any of the prebendaries shall keep a concubine or a chamberwoman, and shall not

vacet modo premisso per ordinarium conferenda. Prepositus vero dicti Collegii quocies dictam preposituram vacare contigerit per nos et successores nostros Reges Scocie ordinario presentabitur. Vicarii ad extra ecclesiarum predictarum per prepositum et capitulum suum et prebendarii dicti Collegii ordinario presentabuntur a quo recipiant canonicam institucionem . et nullus prebendarius instituatur nisi sufficiens fuerit legendo [et] concinendo plano cantu et discantu. Pueri vero dociles ad premissa existant. Volumus eciam et ordinamus quod quilibet prebendariorum cum dispositus fuerit missam celebret qui post missam sacerdotalibus indutus ornamentis cum isopa ad fundatricis sepulchrum accedet ibique *De profundis* cum oracione fidelium et populorum ad deuocionem exortacione deuote perleget . Volumus eciam et ordinamus quod matutina a festo Penthecostes vsque ad festum Sancti Michaelis hora inchoentur quinta de mane et a festo Sancti Michaelis vsque ad Penthecosten hora de mane sexta inicium accipiant et quod statim matutinis finitis ad altare Beate Virginis eadem missa ebdomidalis secundum tabulam pro itinerantibus celebretur et similiter fiat missa ebdomidalis in capella hospitalis hora nona pro pauperibus et infirmis

dismiss her on being thrice admonished by the Provost, his prebend shall be held to be, and shall be vacant and at the collation of the Ordinary in manner aforesaid. Moreover, a Provost of the said College, as often as the said provostry shall happen to be vacant, shall be presented to the Ordinary by us and our successors the Kings of Scotland. The extra-collegiate Vicars of the churches aforesaid, and the Prebendaries of the said College shall be presented by the Provost and his Chapter to the Ordinary, from whom they shall receive canonical institution; and no prebendary shall be appointed unless he shall be capable of reading and singing in plain chant and descant. The boys, moreover, shall be capable of learning these. We will also, and ordain that each of the prebendars, when he shall be disposed, shall celebrate mass, and after mass, robed in his sacerdotal habiliments, shall go with hyssop to the tomb of the foundress, and there shall devoutly read the *De profundis* with the prayer of the faithful and an exhortation of the people to devotion. We will also and ordain that from Whitsunday to Michaelmas matins shall begin at the fifth hour in the morning, and from Michaelmas to Whitsunday at the sixth hour in the morning, and that immediately on matins being concluded, at the altar of the blessed Virgin, the weekly mass, according to the table, shall be celebrated

ibidem degentibus. Volumus eciam et ordinamus vt prepositus et prebendarii antedicti tempore vite nostre anniuersarium pro quondam illustrissimo principe Jacobo Scotorum Rege nostro tenerrimo coniuge deuote impendant et post nostrum decessum anniuersarium suum et nostrum pro nobis et nostris liberis antecessoribus et successoribus nostris die obitus sui et nostri necnon pro prefato reuerendo in Christo patre Jacobo moderno episcopo Sanctiandree post suum decessum perpetuis futuris temporibus decantent et celebrent. Insuper statuimus et ordinamus quod dicti prebendarii in Ecclesia Collegiata antedicta administraturi in suo primo introitu preposito pro tempore existenti obedienciam prestent manualem. Reformaciones et correcciones defectuum circa regimen diuini cultus in ipso collegio contingencium ad prepositum cum cohercione debita pertineant qui eciam defectus in penis pecuniariis seu aliis secundum statuta prepositi et capituli desuper condenda permutari ac leuari valeant et debeant. Postremo statuimus et ordinamus vt nobis facultas omnimoda et plena potestas quoad vixerimus addendi diminuendi et permutandi et singula obscura et minus clara illucidandi in premissis singulis eorum articulis et

for travellers; and similarly a weekly mass shall be performed in the chapel of the Hospital, at the ninth hour, for the poor and infirm living there. We will also, and ordain that the Provost and Prebends aforesaid shall, during our lifetime devoutly observe an anniversary for the late most illustrious prince, James King of Scots, our most tender husband, and, after our decease, shall, on the days of his and our decease, sing and celebrate his and our anniversaries for us and our children, our ancestors, and successors, as also for the foresaid reverend father in Christ, James the present bishop of St Andrews, after his decease in all time to come. Moreover, we statute and ordain that the said Prebendaries, who are to minister in the Collegiate Church aforesaid, shall, on their first entrance, give manual obedience to the Provost for the time being. The reformation and correction of defects occurring in the order of divine worship in the said College shall pertain to the Provost, with due coercion, and these defaults also can and ought to be atoned for and discharged by pecuniary and other penalties according to the statutes to be framed by the Provost and Chapter thereanent. Lastly, we statute and ordain that every kind of faculty and full power be reserved to us so long as we live to add, diminish, and alter, and to elucidate everything obscure and not sufficiently clear, in all the several premises, their

membris incidentibus dependenciis et connexis in vberiori forma secundum iurisperitorum et virorum prudencium consilium reseruetur. Vestram igitur reuerendam paternitatem humiliter precamur et requirimus quatenus dicti Collegii fundacionem dotacionem et diuisionem seu distribucionem per nos vt premittitur factas in omnibus suis punctis membris et articulis per vestram solicitudinem pastoralis officii ad effectum debitum perducendas approbare ratificare et confirmare dignaremini gracia speciali. In cuius rei testimonium magnum sigillum nostrum presentibus est appensum testibus reuerendo in Christo patre Andrea episcopo Glasguensi, venerabili patre Henrico abbate de Pasleto, Andrea domino Avondale cancellario Scocie, Georgio comite Angusie, Alexandro domino Montgomery et Joanne Ros de Halkheid milite, apud Perthe vigesimo quinto die mensis Marcii anno Domini millesimo quadringentesimo sexagesimo secundo.

Post quarumquidem literarum premissarum presentacionem et recepcionem pro parte dicte illustrissime principisse nobis fuit humilter suplicatum quatenus ipsius Collegii fundacionem dotacionem ordinacionem disposicionem diuisionem et distribucionem sic vt premittitur factas in omnibus articulis punctis et membris nostra pastorali auctori-

articles, members, incidents, dependencies, and connections in fuller form, according to the advice of men of legal skill and prudence. We therefore humbly pray and require you, reverend father, that, of special grace, you, of the solicitude of your pastoral office, will be pleased to approve, ratify, and confirm the foundation and endowment of the said College, with the division or distribution made by us as aforesaid in all their points, members, and articles, to be carried to due effect. In testimony of which, our great seal has been affixed to these presents, witnesses, the reverend father in Christ, Andrew Bishop of Glasgow, the venerable father Henry Abbot of Paisley, Andrew Lord Avondale Chancellor of Scotland, George Earl of Angus, Alexander Lord Montgomery, and John Ross of Halkhead, Knight, at Perth, the twenty-fifth day of the month of March, the year of our Lord one thousand four hundred and sixty-two.

After the presentation and reception of which Letters aforesaid, on the part of the said most illustrious Princess, it was humbly supplicated of us that of our pastoral authority we should be pleased to ratify, approve, and confirm the foundation, endowment, ordinance, disposition, division, and distribution of the said

tate ratificare approbare et confirmare dignaremur. Nos vero Jacobus Sanctiandree episcopus supradictus ipsius excellentissime principisse et domine nostre deuotis precibus et piis inclinati desideriis ad Dei Omnipotentis laudem et honorem et perpetuam animarum predictarum memoriam et salutem matura deliberacione et solempni tractatu cum priore et capitulo nostre ecclesie in talibus arduis fieri solitis prehabitis prefatam Collegii Sancte Trinitatis prope Edinburgum ipsius illustrissime principisse fundacionem dotacionem ordinacionem et disposcionem ac reddituum terrarum ecclesiarum et fructuum earundem donaciones concessiones ordinaciones diuisiones distribuciones modo et forma quibus premittitur factas in omnibus suis punctis membris et articulis in vberrima forma que per iurisperitorum prudenciam excogitari poterit pro nobis et successoribus nostris perpetuis futuris temporibus cum consensu et assensu expresso nostri capituli approbamus ratificamus et confirmamus saluis nobis et successoribus nostris obediencia et iurisdiccione ac aliis iuribus ordinariis episcopalibus et archidiaconalibus nobis de iure communi ante huiusmodi annexacionem debitis et consuetis. In cuius rei testimonium sigillum nostrum auctenticum vnacum

College, made as aforesaid in all articles, points, and members. Wherefore we, James, Bishop of St Andrews aforesaid, listening favourably to the devout prayers and pious desires of the said most excellent Princess and our Lady, for the praise and honour of Almighty God, and the perpetual memory and weal of the souls aforesaid, having held mature deliberation and solemn conference with the Prior and Chapter of our Church as is wont to be done in matters of such difficulty, for ourselves and our successors in all time to come, with the express consent and assent of our Chapter, approve, ratify, and confirm the foresaid foundation, endowment, ordination, and disposition, by the said most illustrious Princess, of the College of the Holy Trinity, near Edinburgh, and the donations, grants, ordinances, divisions, and distributions of the revenues, lands, churches, and fruits of the same, made in manner and form as aforesaid, in all their points, members, and articles, and in the fullest form which could be devised by the prudence of persons skilled in law, reserving to us and our successors, the obedience and jurisdiction and other rights ordinary, episcopal and archidiaconal due and wont to pertain to us of common law before such annexation. In testimony whereof our authentic seal, along with the common seal of our chapter, has

sigillo communi capituli nostri presentibus est appensum apud Sanctumandream primo die mensis Aprilis anno Domini millesimo quadringentisimo sexagesimo secundo et consecracionis nostre vigesimo quinto.*

been affixed to these presents, at St Andrews, the first day of the month of April, the year of our Lord one thousand four hundred and sixty-two, and of our consecration the twenty-fifth year.

III.

BULL by Pope Pius II. addressed to Queen Mary of Gueldres, confirming the annexation of the Hospital of Soltray to the Collegiate Church and Hospital of the Holy Trinity. Dated 14 Kal. Julii (18th June) 1462.

PIUS EPISCOPUS etc. Carissime in Christo filie Marie Regine Scotorum Illustri, salutem etc. Singularis devotionis affectus, quem ad nos et Romanam ecclesiam gerere comprobaris, non indigne meretur, ut his, que ad divini nominis laudem et gloriam, pauperum quoque et

PIUS the Bishop, etc. To our most beloved daughter in Christ, Mary, Illustrious Queen of Scots, greeting, etc. The disposition of singular devotion which thou art shewn to bear towards us and the Church of Rome, not unworthily, well deserves that we should add the confirmation of our apostolical muniment to

* [*Note by Sir Lewis Stewart.*—"I extractit this fundatione furth of ane registrie given me in len be Mr Alexander Guthrie towne clerk of Edinburgh and the same is subscryved be Thomas Cottis nottar publict as ane just extract of the originall fundatioune vi November im iiiic lxxxiiii befor Donald Ross dean of Caithnes, Clement Cor, Alexander Elphinstoun and George Newtoun—The nottars subscription hes na signe."]

ogenorum profectum atque animarum salutem pia sunt ordinatione disposita, ut continuis proficiant incrementis ac perpetua illibata persistant, apostolici muniminis adiiciamus firmitatem. Dudum siquidem pro parte tua nobis exposito, quod olim postquam felicis recordationis Nicolaus papa V. predecessor noster, ex certis causis tunc expressis, hospitale pauperum de Soltre Sancti Andree diocesis in unam dignitatem, Cancellariam noncupatam, in ecclesia Sancti Andree de consensu quondam Alani Cant, ipsius hospitalis tunc Rectoris, per suas litteras auctoritate apostolica erexerat, ac de ipsa Cancellaria eidem Alano providerat, Cancellaria predicta, que de iure patronatus Regis Scotorum pro tempore existit, per obitum dicti Alani, qui usque ad ultimum vite sue dictam Cancellariam ipsam pacifice tenuit et possedit, ac tandem extra Romanam Curiam decessit, vacante, clare memorie, Iacobus Scotorum Rex, verus, dum viveret, et unicus ipsius Cancellarie patronus, dilectum filium Iohannem Tyri clericum Sancti Andree diocesis, bacallarium in decretis, ad Cancellariam ipsam sic vacantem Vicario Venerabilis fratris nostri Episcopi Sancti Andree in spiritualibus generali infra tempus legitimum presentaverat, ipseque Vicarius eundem

those arrangements which have been made by pious ordinance for the praise and glory of the Divine name, as well as for the advantage of the poor and necessitous, and the weal of souls, so that they may be enriched by continual additions, and remain for ever unimpaired. Inasmuch as it was explained to us a short time ago on thy behalf, that after our predecessor, Pope Nicholas the fifth of happy memory, for certain explicit reasons at the time had, with apostolic authority, by his letters erected the Hospital for the poor of Soltray, in the diocese of St Andrews, into an office of dignity in the Church of St Andrews, named the Chancellorship, with consent of the late Allan Cant, then rector of the said hospital, and had provided the said Allan with the same chancellorship, the chancellorship foresaid, which belongs by right of patronage to the King of Scotland for the time, becoming vacant by the death of the said Allan, who, to the last day of his life, peaceably held and possessed the said chancellorship, and at length died outwith the Roman Court, James, King of Scots, of famous memory, the true and only patron of the said chancellorship while he lived, had presented, within the lawful time, our beloved son John of Tyre, clerk of the diocese of St Andrews, batchelor-in-laws, for the chancellorship thus vacant, to the Vicar-general in things spiritual of our venerable brother

Iohannem in Cancellarium eiusdem ecclesie ordinaria auctoritate instituerat, quarum quidem presentationis et institutionis vigore idem Iohannes Cancellariam predictam pacifice fuerat assecutus; ac subiuncto, quod tu ad laudem Omnipotentis Dei unam Collegiatam ecclesiam cum hospitali pauperum prope burgum de Edymburg ex parte boreali, dicte diocesis, pro divini cultus augmento, ac Christi pauperum et aliarum miserabilium personarum receptione et sustentatione, de bonis tibi a Deo collatis de novo fundaveras, ac egregio quodam opere construi et edificari feceras, et tam carissimus in Christo filius noster Iacobus Scotorum Rex Illustris, cuius progenitores ipsum hospitale de Soltre ad usum Christi pauperum fundarunt, voluntates suorum progenitorum pro posse observari facere, quam tu, ut prefatum in Cancellariam erectum hospitale ad statum pristinum restitueretur, ipsique novo hospitali erecto incorporaretur, ferventer exoptabatis, prefatus quoque Rex ad hoc consensum prestare paratus existeret, et prout accepimus, ipse Iohannes in favorem dicte unionis, et ut etiam Cancellaria ipsa ad hospitale de Soltre, sicuti prius fuerat, reduceretur, eandem Cancellariam, quam obtinebat, sponte et libere resignare pro-

the Bishop of St Andrews, and the said vicar, by his authority as Ordinary, had inducted the said John to the Chancellorship of the same church, by virtue of which presentation and induction the said John had peaceably obtained the said Chancellorship; and it being subjoined that thou, of the goods bestowed on thee by God, hadst anew founded, and in splendid manner hadst caused to be constructed and erected, to the praise of Almighty God, a Collegiate Church with an Hospital for the poor, near the Burgh of Edinburgh, on the north side, in the said diocese, for the furthering of divine worship, for the reception and maintenance of Christ's poor and other miserable persons; and both our most beloved son in Christ, James, illustrious King of Scots, whose forefathers founded the said Hospital of Soltray for the use of Christ's poor, in order that the wishes of his ancestors might be observed as far as possible, and thyself, fervently desired that the foresaid Hospital, which had been erected into a Chancellorship, should be restored to its pristine state and be incorporated with the same newly erected Hospital for ever, and that the foresaid King was prepared to give his consent to this, and, as we understood, the said John proposed of his own will and freely to resign the said Chancellorship which he held, in favour of such union, and also in

poneret: Nos tuis in ea parte precibus inclinati, Episcopo Glasguensi, eius proprio nomine non expresso, per alias nostras litteras dedimus in mandatis, ut vocatis Episcopo Sancti Andree prefato, ac dilectis filiis Capitulo dicte ecclesie Sancti Andree, et aliis, quorum interest, de premissis omnibus et singulis, eorumque circumstantiis universis auctoritate nostra se diligenter informaret, et si per informationem huiusmodi ita esse reperiret, ab ipso Iohanne vel procuratore suo ad hoc ab eo specialiter constituto resignationem huiusmodi, si illam in suis manibus sponte facere vellet, auctoritate nostra ea vice duntaxat reciperet et admitteret, eaque per ipsum recepta et admissa, Cancellariam ipsam, cum per resignationem ipsam vacaret, dummodo tunc non esset in ea alicui specialiter ius quesitum, ac prefati Regis ad id expressus assensus accederet, nomine dignitateque ipsius Cancellarie in ipsa ecclesia per eum suppressis et penitus extinctis, eodem hospitali de Soltre in statum pristinum reducto, illudque in omnibus iuribus et pertinentiis suis novo hospitali prefato eadem auctoritate perpetuo uniret, annecteret et incorporaret; ita quod liceret ex tunc Rectori dicti novi hospitalis per se, vel alium seu alios corporalem possessionem hospitalis de Soltre, iuriumque et pertinentiarum predictorum auctori-

order that the said Chancellorship should be reconverted into the Hospital of Soltray, as it had been formerly: We, being favourably disposed to thy petition in that regard, by others our letters commissioned the Bishop of Glasgow, his proper name not being expressed, that the Bishop of St Andrews aforesaid, and our beloved sons the Chapter of the said Church of St Andrews, and others whom it concerns, being summoned, he might inform himself diligently on our authority concerning all and sundry the things aforesaid and all their attendant circumstances, and if by such information he should find it so to be, to receive and accept such resignation, by our authority, from the said John or his procurator appointed by him for this purpose, but only in case he should wish freely to make the same at his own hands; and the said resignation being received and accepted by him, to take over the said Chancellorship, when, by such resignation, it should become vacant, provided no one should then have any special right in it, and that the express consent of the said King should be added; and the name and dignity of the same Chancellorship in the said church having been by him totally suppressed and extinguished, and the same Hospital of Soltray being reduced

tate propria apprehendere et perpetuo retinere, ac illius fructus, redditus et proventus in sustentationem pauperum et infirmorum, ac alias in utilitatem ipsorum hospitalium iuxta formam constitutionis pie memorie Clementis pape V. predecessoris super hoc in concilio Viennensi edite convertendos percipere, diocesani loci et cuiusvis alterius super hoc licentia minime requisita, prout in dictis litteris, in quibus quod Alanus Cant tunc ipsius hospitalis de Soltre Rector existeret, et illud in Cancellariam predictam erigi fecisset, expressum fuit, plenius continetur. Cum autem pro parte tua nuper nobis exhibita petitio continebat, [quod] Venerabilis frater noster Thomas modernus Episcopus Dunkeldensis, et non Alanus prefatus, dictam Cancellariam erigi procuraverit, ac tempore date litterarum earundem ecclesia prefata nondum in collegiatam totaliter erecta, neque sumptuosis structuris completa fuerit, ac prius dictus Iohannes resignationem ipsam in manibus prefati Episcopi Sancti Andree, et deinde eiusdem Episcopi Glasguensis fecerit, propter que de ipsarum litterarum viribus posset forsitan hesitari:

to its pristine state, by the same authority, to unite, annex, and incorporate it, in all its rights and pertinents, to the said new Hospital for ever; so that it might be lawful thenceforth to the Rector of the said new Hospital, by himself or others, one or more, on his own authority, to take and retain for ever corporal possession of the Hospital of Soltray and its rights and pertinents aforesaid, and its fruits, rents, and profits, for the sustentation of the poor and infirm, and otherwise for the use of the same Hospitals, according to the form of constitution of our predecessor, Pope Clement V. of pious memory, promulgated regarding this matter in the Council of Vienne, the permission of the diocesan of the place, or of any other whatever, to this effect, not being in any way required, as in the said letters (in which it was expressed that Allan Cant was then rector of the said Hospital of Soltray, and had caused it to be erected into the foresaid Chancellorship) is more fully contained. But since the petition lately presented to us on thy behalf set forth that our venerable brother Thomas, present Bishop of Dunkeld, and not the foresaid Allan, procured the said Chancellorship to be erected, and at the time of the date of the same letters the foresaid church had not as yet been wholly collegiated, nor completed in splendid structure, and that the forementioned John made the said resignation in the hands of the foresaid Bishop of St Andrews, and

Nos ne propter ea dicte littere de surreptione notari valeant, providere volentes, tuis in hac parte supplicationibus inclinati, volumus et apostolica auctoritate decrevimus, quod littere nostre predicte ac processus habiti per easdem, et quecunque inde secuta a data presentium valeant, plenamque roboris firmitatem obtineant in omnibus et per omnia, ac si in illis, quod Thomas et non Alanus erectionem predictam fieri procurasset, ac Iohannes prefatus dictam resignationem in manibus prefati Episcopi Sancti Andree loci ordinarii primo ad effectum huiusmodi, et subsequenter in ipsius Glasguensis Episcopi manibus fecisset, expressum extitisset, quodque tempore litterarum nostrarum earundem dicta ecclesia totaliter expleta et in collegiatam ecclesiam erecta foret, nulla mentio facta fuisset. Non obstantibus premissis, ac omnibus illis, que in dictis litteris non obstare voluimus, ceterisque contrariis quibuscunque. Nulli ergo etc. nostre voluntatis et constitutionis infringere etc. Si quis autem etc. Datum Viterbii Anno etc. M.CCCC.LXII. quartodecimo Kal. Iulii, Pontificatus nostri anno quarto.

thereafter in those of the same Bishop of Glasgow, on account of which doubts might perchance be entertained regarding the validity of the said letters, We, being favourably disposed towards thy supplications in that regard, desiring to guard against the possibility of the said letters being on that account branded as surreptitious, will, and by apostolic authority have decreed, that our foresaid letters, and the proceedings had in respect thereof, and whatsoever has followed thereon, may from the date of these presents have force, and may obtain the full force of validity in all and through all, as if in them it had been expressly set forth that Thomas, and not Alan, had procured the said erection to be made, and that the foresaid John had made the said resignation to this effect, first in the hands of the foresaid Bishop of St Andrews, the Ordinary of the See, and afterwards in those of the Bishop of Glasgow, and that at the time of our said letters, no mention had been made that the said church was wholly complete, and had been erected into a Collegiate Church. The premises, and all those things in the said letters which we wish not to stand in the way, and everthing else to a contrary purpose notwithstanding. Let no one, therefore (dare) to infringe (the tenor) of our will and constitution. But if any one, etc. Given at Viterbo in the year etc. M.CCCC.LXII. the fourteenth calend of July, of our pontificate the fourth year.

IV.

BULL by Pope Pius the Second reciting the foundation by Queen Mary of Gueldres of the Collegiate Church and Hospital of the Holy Trinity, and confirming the annexation thereto of the Hospital of Soltray, and the Chapel of Utherogall in Fife. Dated 6th Idus Julii (10th July) 1462.

PIUS EPISCOPUS, etc. Ad futuram rei memoriam. Inter multiplices curas que nobis ex Apostolatus officio censentur incumbere, illas libenter amplectimur per quas in ecclesiis quibuslibet devote solicitudinis studio benedicatur Altissimus ac in hospitalibus et aliis piis locis pauperum et aliarum miserabilium personarum necessitatibus succuratur, nec non hiis que propterea processisse comperimus ut perpetuo illibata persistant Apostolici muniminis adjicimus firmitatem. Sane pro parte dilecte in Christo filie nostre Marie Regine Scotorum illustris nobis nuper exhibita petitio continebat, quod olim ipsa de propria salute recogitans et ad caritatis opera maxime circa egenos et pauperes manus juxta evangelica documenta extendens, dum adhuc clare memorie Jacobus II. Rex Scotorum vir ejus ageret in humanis, quoddam Hospitale pro susceptione

PIUS the Bishop, etc. For future memory of the fact. Among the manifold cares which are deemed to be incumbent on us, by reason of our apostolic office, we willingly embrace those by which, in all churches, with the zeal of devout solicitude, the Most High is praised and succour rendered to the necessities of poor and other wretched persons in Hospitals and other pious places, and to those foundations which we find have made progress in this respect, that they may remain unimpaired for ever, we have added the strength of apostolic confirmation. Truly the petition, lately presented to us on the part of our beloved daughter in Christ, Mary the illustrious Queen of Scots, contained, that formerly she, having regard to her own weal, and for works of charity, especially to the poor and needy, stretching out her hands according to the lessons of the Gospel, whilst as yet James II. of distinguished memory, King of Scots her husband was alive,

pauperum et egenorum eorundem extra oppidum regium de Edinburgh Sancti Andree diocesis construi et edificari fecit, illudque dotavit et successive boni operis fructum provida consideratione attendens et ad majora pietatis officia ferventius aspirans juxta Hospitale predictum quandam insignem ecclesiam sive capellam pro Collegio fundato unius videlicet prepositi et decem aut duodecim presbyterorum et clericorum inibi collegialiter et perpetuo Altissimo serviturorum erexit, ac ecclesiam predictam magnifico et sumptuoso opere continuavit, et quantotius illam divina cooperante gratia perficere intendit, nec non quandam bonam baroniam atque alias possessiones census annuos redditus et bona usque ad summam centum mercarum vel circiter in dotem eidem ecclesie assignavit et annuente Domino multo amplius in posterum dotem ipsam ampliare intendit. Cum autem felicis recordationis Nicolaus Papa V. predecessor noster quoddam Hospitale pauperum de Soltre nuncupatum dicte diocesis, quod de jure patronatus Regis Scotorum pro tempore existentis fore dignoscitur, et deformi ruine subjacebat, et in quo nulla hospitalitas servabatur, in Cancellariam ecclesie Sancti Andree, per quas-

caused to be constructed and built a certain Hospital for the reception of poor and needy persons outwith the royal City of Edinburgh, in the diocese of St Andrews, and endowed the same, and afterwards waiting with provident consideration for the fruit of good works, and more fervently aspiring to greater offices of piety, she erected near the Hospital aforesaid a noble church or chapel for the College that had been founded, viz., of one Provost and ten or twelve priests and clerks there in collegiate manner, and perpetually to serve the Most High, and has continued the church aforesaid with magnificent and sumptuous work, and intends as soon as possible, with the assistance of divine grace, to finish the same. She has besides assigned for the endowment of the said church a good barony and other possessions, property, annual rents, and goods to the sum of one hundred merks or thereby; and with the favour of God intends more amply to enlarge the said endowment in future. Since, moreover, our predecessor, Pope Nicholas V. of happy memory, by certain letters of his, ordered a certain hospital for the poor, called the Hospital of Soltray, of said diocese, which in right of patronage is known to belong to the King of Scots for the time being, and lay in shapeless ruin, and in which no hospitality was kept, to be erected into a Chancellorship of the church of St

dam suas [literas] erigi, primo et deinde nos ex certis rationabilibus causis tunc expressis, Hospitale predictum quod de jure patronatus prefati Regis existit, ad humiles supplicationes ipsius Regine, rescissa unione et erectione Cancellarie hujusmodi, novo Hospitali prefato per alias nostras literas uniri mandavimus, prout in singulis literis predictis ac super illis decretis processibus dicitur plenius contineri: Pro parte carissimi in Christo filii nostri Jacobi moderni Regis Scotorum, ipsius Regine primogeniti, ac etiam prefate Regine, nobis fuit humiliter supplicatum ut fundationi dotationi annexationi et unioni predictis, ac etiam capelle de Otherroga[ll]e dicte diocesis, que similiter de jure patronatus prefati Regis existit, et cujus ac Hospitalis de Soltre predicti fructus redditus et proventus septuaginta librarum sterlingorum secundum communem estimationem valorem annuum ut asseritur non excedunt, eidem Hospitali ordinaria auctoritate unire et pro eorum subsistentia firmiori robur Apostolice confirmationis adjicere nec non ad abundantem cautelam ecclesiam predictam in Collegiatam cum insigniis Collegialibus de novo erigere, et alias super hiis oportune providere de

Andrews, and we afterwards, in answer to the humble supplications of the foresaid Queen, for certain reasonable causes then expressed, by our other letters commanded the Hospital aforesaid, which in right of patronage belongs to the foresaid King, the union and erection of such Chancellorship being annulled, to be united to the new Hospital aforesaid, as in the several letters aforesaid, and the decrees and processes regarding them is said to be more fully contained. It was humbly supplicated of us, on the part of our most dearly beloved son in Christ, James the present King of Scots, the eldest son of the said Queen, and also of the foresaid Queen, that we should of our apostolic benignity vouchsafe to confirm the foundation, endowment, annexation, and union aforesaid, and to unite to the same Hospital by authority of the Ordinary, the chapel of Utherogall of said diocese, which in like manner belongs in right of patronage to the said King, and of which, and of the Hospital of Soltray, the foresaid fruits, revenues, and proceeds do not, as is asserted, exceed the annual value of seventy pounds sterling, according to the common estimate, and also, for the more secure subsistence of these foundations to add the strength of apostolic confirmation, and also, for abundant security, of new to erect the foresaid church into a Collegiate Church with collegiate insignia, and otherwise to make suitable

benignitate Apostolica dignaremur. Nos igitur, qui divini cultus augmentum ac pauperum et miserabilium personarum commoda tota mente appetimus, hujusmodi supplicationibus inclinati, fundationem dotationem uniones et annexationes predictas ratas, habentes et gratas eas et quecunque inde secuta auctoritate Apostolica, tenore presentium, approbamus et confirmamus, supplentes omnes defectus, si qui forsan intervenerint in eisdem, et nichilominus pro potioris cauthele suffragio eandem ecclesiam in Collegiatam cum insignibus Collegialibus de novo erigentes, necnon quascunque oblationes fidelium que eidem ecclesie pro tempore obvenient in ejus fabricam et reparationem committendas fore, decernentes ipsius novi Hospitalis magistro sive rectori moderno et qui pro tempore erit quod univeris et singulis utriusque sexus fidelibus in eo pro tempore commorantibus et decedentibus in eodem nec non aliis in illo servientibus omnia et singula ecclesiastica sacramenta ministrare libere et licite valeat plenam et liberam auctoritate presentium potestatem concedimus et facultatem, Non obstantibus constitutionibus et ordinationibus Apostolicis ceterisque contrariis quibuscunque. Nos enim exnunc irritum decernimus et inane si secus super hiis a quoquam

provision regarding these matters. We, therefore, who, with our whole mind, desire the increase of divine worship, and the advantage of poor and miserable persons, listening favourably to these supplications, of apostolic authority, by the tenor of these presents, approve and confirm the foundation, endowment, union, and annexation aforesaid, holding them, and whatsoever has followed thereupon, fixed and approved, supplying all defects, if any perchance shall be found in the same, and nevertheless for the purpose of greater security, erecting of new the said church into a Collegiate Church, with collegiate insignia. Decreeing, moreover, that whatever offerings of the faithful shall for the time fall to the said church, shall be applied upon its fabric and the repair thereof; and we, by the authority of these presents, grant full and free power and authority to the Master of the new Hospital, or to the Rector for the time, to administer all and each of the sacraments of the Church, to all the faithful of both sexes, dwelling therein for a time, and those dying there, as well as to others serving therein, notwithstanding any apostolic constitution, ordinances and others contrary whatsoever. For we from this time forth declare whatever may be attempted to the contrary in these matters, by any one, on any authority, knowingly or

quavis auctoritate scienter vel ignoranter contigerit attemptari. Nulli ergo etc. nostre confirmationis approbationis suppletionis erectionis constitutionis et voluntatis infringere. Si quis autem etc. Datum in Abbatia Sancti Salvatoris Clusinensis diocesis Anno etc. MCCCCLXII. Sexto Idus Julii. Pontificatus nostri anno Quarto.

ignorantly, null and void. Let no one, therefore, etc., dare to infringe the tenor of our confirmation, approbation, restoration, erection, constitution, and will. If any one, moreover, etc. Given in the Abbey of St Salvator, in the diocese of Clusium, the year, etc. MCCCCLXII. sixth of the Ides of July. Of our pontificate the fourth year.

V.

Bull by Pope Pius the Second, granting a plenary indulgence to all who in a devout spirit of contrition visit the Church of Trinity College in the course of five years during the Feast of its Dedication, on the 10th of July or its Octaves, &c. Dated 27th August 1463.

Pius II. etc.—Universis Christi fidelibus presentes literas inspecturis Salutem etc. Maximum certe munus Christiano populo conferri solet cum ei ad diluendas mentium maculas et celestem beatitudinem consequendam opportuna pietas misericorditer se offert presertim ubi simul et domus Dei amplietur et indigentie pauperum Christi succuratur;

Pius II. etc. To all the faithful in Christ, who shall see these present letters, greeting, etc. A very great gift certainly is wont to be conferred on a Christian people, when an opportune piety comes forward compassionately to wash away the stains of souls, and to procure celestial happiness, especially when at the same time both the house of God is enlarged and succour is afforded to the indigence of

quippe vero Omnipotens ipse multo ampliora meritis munera indesinenter ipsis fidelibus retribuit, tamen ad eorum confirmandos excitandosque animos benignum nostrum et Apostolice Sedis pie matris gremium clementer patefacere consuevimus. Igitur intellecto quod per charissimam in Christo filiam Mariam clare memorie Jacobi Secundi Scotorum Regis relictam viduam insignis Collegiata ecclesia sive capella aut Hospitale pauperum Sancte Trinitatis extra oppidum regium de Edinburgh Sancti Andree diocesis fundata, et pro vno preposito ac decem vel duodecim personis ecclesiasticis presbyteris et clericis inibi collegialiter et perpetuo Altissimo famulantibus erecta ac magnifico et sumptuoso opere continuata extitit, et quam seu quod divina cooperante gratia carissimus in Christo filius noster Jacobus tertius etiam Scotorum modernus Rex illustris ejusdem Jacobi Secundi et Marie prefate natus perficere intendit; Nos etiam tam Jacobi tertii quam Marie illustris genetricis sue predictorum consideratione ac ut ipsi fideles ad hujusmodi devotionem et subsidia prestanda promptius inducantur atque inflammentur quo ex hoc ibidem dono celestis gratie

Christ's poor; but although the Almighty himself incessantly repays the faithful with gifts much greater than their merits, yet for confirming and stimulating their minds, we have in our clemency been wont to open our benignant bosom and that of our pious mother the Apostolic See. Having, therefore, learned that a famous Collegiate Church, or Chapel, or Hospital of poor of the Holy Trinity outwith the royal town of Edinburgh, in the diocese of St Andrews, has been founded by our dearest daughter in Christ Mary, the widow of James the Second of renowned memory, King of Scots, and has been erected and carried on by magnificent and expensive workmanship for one provost and ten or twelve ecclesiastics, presbyters, and clerks, who there in a collegiate manner and perpetually serve the Most High, and which, with the assistance of divine grace, our most dearly beloved son in Christ, James the Third, the present illustrious King of Scots, son of the said James the Second and of the foresaid Mary, intends to complete. We also—in consideration both of James the Third and of his illustrious mother Mary aforesaid, and that the faithful may be the more readily induced and incited to similar devotion, and to render assistance, when, from this gift of divine grace there made, they have seen themselves more abundantly

uberius conspexerint se refectos de Omnipotentis Dei misericordia ac beatorum Petri et Pauli apostolorum ejus auctoritate confisi, universis utriusque sexus undecumque existentibus vere penitentibus et confessis Christi fidelibus qui decima die mensis Julii qua dedicatio ipsius ecclesie existit et illius die octava, a primis vesperis usque ad secundas vesperas inclusive, hujusmodi ecclesiam capellam aut Hospitale corde contriti et ore confessi in prefixo quinquennio visitaverint atque manus adjutrices huic pio operi pro modo facultatum suarum et de consilio confessoris sui existentis ex deputatis ut infra dicitur porrexerint, ac eis qui legitime impediti visitationem hanc ut voluissent facere non potuerint, necnon pauperibus Christi fidelibus in eadem ecclesia sive Hospitali pro tempore degentibus qui inibi decesserint, in mortis articulo alias etiam corde contritis et quantum recordarentur ore confessis, omnium peccatorum criminum et excessuum suorum etiam in casibus Sedi Apostolice reservatis plenariam absolutionem ac indulgentiam tenore presentium concedimus et elargimur; presentibus in die dicte dedicationis anni proximi futuri primum inchoaturis ac ut deinceps sequitur de quinquennio in

refreshed, being assured of the mercy of Almighty God, and the warrant of St Peter and St Paul his apostles—by the tenor of these presents, grant and bestow on all the faithful in Christ of both sexes whencesoever they may come, being truly penitent and making confession, who, on the tenth day of the month of July, which is the day of dedication of the said church, and on the octave of the same, from the first vespers to the second vespers inclusive, shall, within the appointed five years, have visited the said church, chapel, or hospital, contrite in heart and confessing with the mouth, and shall have stretched out helping hands to this pious work according to their abilities, and by advice of their confessor, being one of those deputed as after-mentioned; and to those who shall have been lawfully prevented from making this visitation as they had wished to do; likewise to the poor faithful in Christ dwelling in the said Church or Hospital for the time, who have died there, and who at the moment of death, and at other times, were contrite in heart, and confessed with the mouth to the best of their remembrance, plenary absolution and indulgence of all their sins, crimes, and excesses, even in cases reserved to the Holy See; these presents first to come into effect on the day of the said dedication in the year next ensuing, and to continue in force on its successive recurrence every fifth year for the space of

quinquennium usque quinquaginta annos tantum duraturis. Insuper quoque ut dicti fideles sicuti prediximus visitantes et porrigentes facilius animarum suarum salutem et divinam misericordiam consequi valeant, confessoribus ydoneis per dilectum filium prepositum dicte ecclesie et collectorem Apostolice Camere pro tempore existentes juxta confluentium fidelium concursum prout expediens videbitur deputandis secularibus vel quorumcunque ordinum regularibus, in dicta quinquennali die dedicationis hujusmodi quatuor diebus ante et quatuor post et in octavis eorum, confessionibus diligenter auditis, pro commissis debitam absolutionem impendendi et penitentiam salutarem injungendi plenam et liberam etiam tenore presentium concedimus facultatem. Volumus autem quod pecunie quas pro indulgentia hujusmodi consequenti per fideles ipsis erogari contigerit in una capsa duabus clavibus claudenda fideliter et vigilanter conserventur, ipsarumque clavium unam prepositus prefatus et duo seniores ejusdem Collegii alteram vero dicte Camere in regno Scocie collector pro tempore existentes tenere debeant, omniumque pecuniarum et rerum ex hujusmodi oblationibus et elemosinis provenientium tertia pars integra et sine ulla fraude pro defensione et

fifty years only. Moreover, that the said faithful, as we have said above, visiting and assisting, may the more easily obtain the salvation of their souls and the divine compassion, we also, by the tenor of these presents, grant to suitable confessors, seculars or regulars of whatsoever order, to be deputed, as shall seem expedient, by our beloved son the Provost of the said church, and the Collector of the Apostolic Chamber for the time being, according to the concourse of the faithful flowing thither, on the said quinquennial day of the said dedication, four days before and four days after, and on their octaves, after carefully hearing their confessions, full and free power of granting due absolution for sins, and of enjoining salutary penance. We will, moreover, that the moneys which for such subsequent indulgence may be paid to them by the faithful, shall be faithfully and vigilantly kept in a box, to be locked with two keys, and of the said keys the Provost aforesaid and two seniors of the said college for the time being should keep one, and the Collector of the said Chamber in the kingdom of Scotland the other, and of all moneys and effects arising from such offerings and alms, a third part, entire and without any

augmento fidei Catholice adversus impiissimos efferatissimosque Christi nominis inimicos eidem Apostolice Camere reservetur et continuo per eundem collectorem fidelissime efficacissimeque percipiatur reddatur et consignetur, relique autem due tertie per prepositum et duos seniores antedictos in hujusmodi perficienda edificia supplenda ornamenta muniendumque et manutenendum Collegium pauperesque atque locum hujusmodi dumtaxat et non aliter quoquo modo convertantur; alioquin collector prepositus et seniores ipsi duo qui quod absit secus facere aut oblationes elemosinas ac alia hujusmodi in alios usus aut aliter quoquomodo disturbare vel agere presumpserint indignationem ejusdem Omnipotentis Dei et Apostolorum Petri et Pauli predictorum eo ipso incurrant Nulli ergo etc. nostre concessionis elargitionis et voluntatis infringere etc. Si quis autem hoc attemptare presumpserit indignationem ipsius Omnipotentis Dei et Apostolorum Petri et Pauli predictorum se noverit incursurum Datum Tibur anno etc. MCCCCLXIII. Sexto Kalendas Septembris Pontificatus nostri anno Quinto.

fraud, shall be reserved to the same Apostolic Chamber, and shall immediately be most faithfully and effectually taken possession of, handed over, and consigned thereto by the said Collector, for the defence and furtherance of the Catholic faith against the most impious and fierce enemies of the name of Christ; but the two remaining thirds shall be applied by the Provost and two seniors aforesaid for the completion of the said buildings, for supplying ornaments, for repairing and upholding the College, and the poor and said place only, and not in any other way; otherwise the collector, provost, and said two seniors, who shall presume, which heaven forbid, to pervert or use the offerings, alms, and other things of this sort for other purposes, or in any other way whatsoever, shall by such deed incur the indignation of Almighty God, and of the apostles Peter and Paul aforesaid. Let no one, therefore, dare to infringe the tenor of our concession, grant, and will, etc. But if any one shall presume to attempt this, let him know that he will incur the indignation of Almighty God himself, and of the apostles Peter and Paul aforesaid. Given at Tibur the year, etc. MCCCCLXIII. the sixth of the Kalends of September, of our pontificate the fifth year.

VI.

TRANSUMPT, on 21st March 1525–6, of CHARTER by James Archbishop of St Andrews, annexing the Church of Dunnotter to Trinity College, for the support of two Prebendaries, to be called respectively "the Dean" and "the Prebendary of Dunnotter." Edinburgh, 14th November 1502.

VNIUERSIS et singulis Sancte Matris Ecclesie filiis presentes literas inspecturis lecturis visuris pariterque audituris: Jacobus Symsoun, rector de Kyrkforthir ac officialis Sanctiandree principalis salutem in Domino sempiternam: Noueritis quod nos ad instanciam venerabilis viri magistri Alexandri Kynnynmond procuratoris specialiter constituti pro venerabili et egregio viro Domino Johanne Dingwaill preposito Ecclesie Collegiate Sancte Trinitatis prope Edinburgh Sanctiandree diocesis, omnes et singulos interesse habentes seu habere putantes, et quos infrascriptum tangit negotium seu tangere poterit quomodolibet in futurum, ad comparendum coram nobis seu commissariis nostris, pluribus aut vno, certis die et loco inferius designatis, ad videndas et audiendas quasdam literas recolende memorie ac quondam reuerendissimi in Cristo patris et

TO all and sundry sons of Holy Mother Church who shall examine, read, see, and likewise hear these present letters, James Symsoun, rector of Kyrkforthir, and Official principal of St Andrews, eternal salvation in the Lord: Know ye that we, at the instance of the venerable man Master Alexander Kynnynmond, procurator specially constituted for a venerable and worthy man Sir John Dingwall, provost of the Collegiate Church of the Holy Trinity near Edinburgh, in the diocese of St Andrews, have caused to be summoned by public edict, affixed to the walls of the Metropolitan Church of St Andrews, all and sundry having, or thinking they have interest, and whom the matter hereafter mentioned affects, or may in any way affect in time to come, to appear before us or our commissaries, one or more, on a certain day and place hereinafter set forth, to see and hear certain letters—of the late most reverend father in Christ, and lord of venerated memory, James by divine

domini, Jacobi miseratione diuina Sanctiandree Archiepiscopi totius regni Scotie primatis et apostolice sedis legati nati Ducis Rossie ac commendatarii perpetui monasterii de Dunfermling, in originalibus pergamino scriptas, eius sigillo rotundo roboratas, cum consensu et assensu venerabilium et religiosorum virorum, dominorum prioris ac conuentus Sanctiandree, capitulariter congregatorum, eorum sigillo capitulari munitas, ac signo et subscriptione quondam circumspecti viri magistri Patricii Middilton notarii publici etiam subscriptas et roboratas, de et super erectione creatione et deputatione ecclesie perrochialis de Donnottir dicte Sanctiandree diocesis, tam rectorie quam vicarie eiusdem, ac vnione annexatione et incorporatione eiusdem ecclesie per dictum reuerendissimum in duabis prebendis pro duobis prebendariis perpetuo Deo seruituris in ecclesia Collegiata Sancte Trinitatis prope Edinburgh antedicta, prout in hujusmodi literis erectionis desuper confectis et conscriptis in libro regestri monasterii Sanctiandree, et ibidem ad perpetuam rei memoriam registratis, plenius continetur, transsumi copiari publicari et exemplari ac in publicam transsumpti formam redigi, aliaque fieri et executioni demandari, que in hac parte de jure requiritur, in valuis

mercy Archbishop of St Andrews, primate of the whole kingdom of Scotland, and legate natus of the apostolic see, duke of Ross, and perpetual commendator of the monastery of Dunfermling, written in the original on parchment, confirmed by his seal, with the consent and assent of venerable and religious men, the lord prior and convent of St Andrews assembled in chapter, sealed with the chapter seal, and subscribed and affirmed with the sign and subscription of the late circumspect man Master Patrick Middilton, notary public, of and concerning the erection, creation, and deputation of the parish church of Dunnotter in the said diocese of St Andrews, as well parsonage as vicarage of the same, and of the union, annexation, and incorporation of the same church, by the said most reverend, into two prebends for two prebendaries to serve God for ever in the foresaid Collegiate Church of the Holy Trinity near Edinburgh, as in the same letter of erection made thereupon and engrossed in the register book of the monastery of St Andrews, and registered there for perpetual memory of the fact, is more fully contained,—transumed, copied, published, and drawn up in the the form of a public transumpt, and other things done and demanded to be done as in this part is required by law; On which

ecclesie metropolitane Sanctiandree per edictum publicum ibidem affixum, citari fecimus: Quo die aduenicnte et comparente coram nobis dicto procuratore prefati domini prepositi et huiusmodi librum regestri prefati monasterii Sanctiandree coram nobis judicialiter exhibente et vocatis interesse habentibus et non comparentibus, ipsis contumacibus reputatis, huiusmodi librum regestri, necnon prefatam erectionem coram nobis collationatam in eodem expressatam, cum originali de uerbo in uerbum concordantem, per testes juratos sufficienter in forma juris recognitam, transsumi, copiari, publicari, et per notarium publicum subscriptum exemplari, et in formam juris redigi decreuimus, cum omni juris solemnitate, quod presenti transsumpto siue instrumento publico, erectionem antedictam in se continenti, tanta fides in judicio et extra ac vbique locorum de cetero detur et adhibeatur qualis et quanta prefate erectioni originali daretur et adhiberetur si in medium produceretur.

Sequitur tenor prefate erectionis dicte Ecclesie de Donnottir, cum suis pertinentiis, prout in huiusmodi libro regestri dicti monasterii Sanctiandree registratur et continetur, de quo supra fuit mentio.

JACOBUS miseratione diuina Sanctiandree Archiepiscopus, totius regni

day, compearing before us the said procurator of the foresaid Provost, and exhibiting before us judicially the said register book of the foresaid monastery of St Andrews, and those having interest having been cited, and not compearing, being reckoned disobedient, we decerned with all solemnity of law that the said register book, and the foresaid erection contained in the same, having been collated before us, and acknowledged by witnesses duly sworn in legal form as agreeing with the original word for word, should be transumed, copied, published, and transcribed, and drawn up in legal form, by the underwritten notary public, that as much credit may be given and accorded, both in and out of court and everywhere else, to the present transumpt or public instrument, containing in it the foresaid erection, as would be given and accorded to the original erection if it were publicly produced.

Here follows the tenor of the foresaid erection of the said church of Dunnotter with its pertinents, as registered and contained in the said register book of the monastery of St Andrews, of which mention has been above made:—

JAMES by divine mercy Archbishop of St Andrews, Primate of the whole

Scotie primas et apostolice sedis legatus, Dux Rossie, et commendatarius perpetuus monasterii de Dunfermling etc. vniuersis sancte matris Ecclesie filiis ad quorum notitias presentes litere peruenerint, salutem in omnium saluatore. Ut ecclesiarum omnium et presertim collegiatarum per prouinciam et diocesim nostram consistentium, et ad illarum decorem inibi dignitates obtinentium personarum status salubriter dirigi seruarique possit honestius, ac numerus personarum diuinum ibidem psallentium officium oportunior habeatur, nostri libenter fauoris impertimur presidium, potissime cum temporum requirit necessitas, cause persuadent rationabiles, et diuini cultus augmentum salubriter id exposcit: Exhibita siquidem nobis nuper per dilectum nostrum clericum magistrum Johannem Brady prepositum Ecclesie Collegiate Sancte Trinitatis prope Edinburgh nostre diocesis, petitio continebat, quod in dicto Collegio tantum octo prebendarii pro nunc existunt, et si numerus prebendariorum augeretur, in non modicum diuini cultus augmentum et ipsius ecclesie collegiate honorem et vtilitatem cederet et deueniret, et sicut eadem petitio subiungebat si parochialis ecclesia de Donnottir nostre diocesis, ad nostram collationem et plenariam dispositionem spectans et

kingdom of Scotland, and Legate of the apostolic see, Duke of Ross, and perpetual commendator of the monastery of Dunfermling, etc. : To all the sons of Holy Mother Church to whose notice these present letters shall come, greeting in the Saviour of all: We willingly grant our friendly aid, especially when the necessity of the times requires, reasonable causes induce, and the salutary advancement of divine worship demands, that the condition of all churches, especially collegiate, throughout our province and diocese, and for their honour, of persons holding dignities therein, may be wholesomely regulated, and more creditably maintained, and that the number of persons chanting divine service in the same may be more adequate; inasmuch as a petition lately presented to us by our beloved clerk, Master John Brady, provost of the Collegiate Church of the Holy Trinity near Edinburgh, in our diocese, set forth that in the said College there are now only eight prebendaries, and that if the number of prebendaries were increased, it would tend to the great advancement of divine worship, and to the honour and profit of the Collegiate Church itself, and as the same petition went on to say, if the parish church of Dunnottir in our diocese, belonging and pertaining to our collation and free grant, were

pertinens, dicto Collegio vniretur et incorporaretur, dueque prebende pro duobis prebendariis ex fructibus dicte ecclesie parochialis, in augmentum dictorum octo prebendariarum inibi existentium constituerentur, crearentur et erigerentur, ipsi ecclesie collegiate plurimum prouideretur, cultusque diuinus exinde augmentaretur in eadem; quare per dictum magistrum Johannem nobis fuit humiliter supplicatum ut super hiis prouidere de ordinaria autoritate dignaremur: Nos vero attendentes petitionem huiusmodi justam et rationi consonam, de premissis omnibus et singulis ac eorum circumstantiis vniuersis inquisiuimus, et nos diligenter informauimus, et per informationem eandem reperimus omnia et singula per dictum magistrum Johannem asserta et narrata, ut premittitur, vera et veritate fulciri: Idcirco autoritate nostra ordinaria qua fungimur in hac [parte] matura deliberatione et solempni tractatu cum priore et capitulo nostre ecclesie in talibus fieri solito prehabitis, ad Dei laudem et Sancte et Indiuidue Trinitatis gloriam et honorem, et in diuini cultus augmentum, duas prebendas pro duobis prebendariis perpetuo Deo seruituris in eadem ecclesia Collegiata Sancte Trinitatis, de

united to and incorporated with the said College, and two prebends for two prebendaries were constituted, created, and erected out of the fruits of the said parish church, in addition to the said eight prebendaries which are now therein, the said Collegiate Church would be well provided for, and divine worship in the same advanced thereby; wherefore we were humbly besought by the said Master John that we would deign of our authority as Ordinary to provide in regard to these matters. And we, considering the said petition to be just and agreeable to reason, have diligently inquired and informed ourselves concerning all and singular the premises and their circumstances, and by the same information have found all and singular the things so asserted and narrated by the foresaid Master John to be true and supported by the truth: Therefore by our authority as Ordinary, which we exercise in these parts, after mature deliberation and solemn consultation previously had with the prior and chapter of our church, as is wont in such cases, to the praise of God and the glory and honour of the Holy and Undivided Trinity, and for the advancement of divine worship, we have created, erected, and deputed, as by these presents we create, erect, and depute two prebends for two prebendaries, to be described below, to serve God for ever in the said Collegiate Church of the Holy Trinity out of the fruits of the said parish

fructibus dicte ecclesie parochialis de Dounottir per nos inferius designandis, creauimus ereximus et deputauimus, prout per presentes creamus erigimus et deputamus, ac dictam ecclesiam parrochialem de Donnottir, cum omnibus juribus et pertinentiis suis, imperpetuum, dictis duobis prebendariis pro suis prebendis, vniuimus annexinus et incorporauimus, ac exnunc vnimus annectimus et incorporamus; ita quod cedente vel decedente Magistro Valtero Stratoun moderno ipsius parochialis ecclesie rectore, seu illam aliis quomodolibet demittente, etiam si actu quouismodo nunc valet, liceat dictis duobis prebendariis corporalem realem et actualem possessionem dicte parrochalis ecclesie jurium et pertinentiarum eiusdem autoritate propria apprehendere, ac in suos dictarum prebendarum ac vicarii per nos creandi vsus et vtilitatem conuertere et perpetuo retinere, alterius superioris licentia super hoc minime requisita: Primus vero istorum prebendariorum vocabitur Decanus, qui in absencia prepositi erit principalis et presidens in choro ac capitulo eiusdem ecclesie, habebitque jurisdictionem iis, prepositi totalem ipso absente, preposito vero presente cessabit omnis jurisdictis et potestas dicti decani; sed primus et principalis post prepositum in ecclesia ac capitulo

church of Dunnottir; and we have united, annexed, and incorporated, and now unite, annex, and incorporate the said parish church of Dunnottir, with all its rights and pertinents for ever to the said two prebendaries for their prebends, so that upon the resignation or decease of Master Walter Stratoun, the present rector of the said parish church, or his demitting it to others in any way, even if he now has any power to do so, it shall be lawful to the said two prebendaries to take corporal, real, and actual possession of the said parish church, and of the rights and pertinents of the same, by their proper authority, and to convert and for ever retain the same for the use and profit of the said prebendaries themselves and of a vicar to be appointed by us, the licence of any other superior being nowise required for that effect. Moreover, the first of these prebendaries shall be called the Dean, who, in the absence of the provost, shall be principal and president in the choir and chapter of the same church, and shall have all the jurisdiction of the provost in the same, the provost himself being absent, in his presence however the whole jurisdiction and power of the said Dean shall cease, but he shall have the first and principal place after the provost in the church and

inter ceteros prebendarios habeatur, habebitque pro sua sustentatione medietatem omnium et singulorum fructuum oblationum et decimarum tam rectorie quod vicarie dicte ecclesie de Dunnottir vnite. Secundus vero prebendarius vocabitur Prebendarius de Dunnottir, eritque expertus ac bene eruditus in organis et ludet in illis festis congruentibus, qui celibrabit pro Jacobo moderno Archiepiscopo Sanctiandree fratre germane Regis Scocie, missam cum dispositus fuerit, in vndecima hora ante meridiem, habebitque pro sua sustentatione aliam dimedietatem fructuum decimarum et oblacionum vicarie et rectorie eiusdem ecclesie de Dunnottir. Et ambo isti prebendarii erunt sacerdotes ac docti et experti in legendo et construendo plano cantu preket not et discantu, et faciant obedientiam proposito in primo introitu jurabuntque seruare statuta ac omnia et singula in primeua fundacione dicte Ecclesie Collegiate ordinata, ac personalem et continuam facient residentiam apud dictam ecclesiam collegiatam per seipsos et non per alios, sicuti ceteri prebendarii dicte ecclesie. Et ne dicta nostra parrochialis ecclesia de Dunnottir propter presentem nostram erectionem et vnionem

chapter among the other prebendaries; and shall have for his maintenance the half of all and singular the fruits, oblations, and teinds, as well parsonage as vicarage of the said united church of Dunnottir. The second prebendary shall be called the prebendary of Dunnottir, and shall be expert and well learned in organs, and shall play upon them on the proper feasts, and he shall celebrate mass for James present Archbishop of St Andrews, brother-german of the King of Scotland, when he shall be appointed, at the eleventh hour before noon, and shall have for his maintenance the other half of the fruits, teinds, and oblations, parsonage and vicarage of the said church of Dunnottir. And both the said prebendaries shall be priests, and learned and expert in reading and understanding plain chant, "preket not" and responding, and shall make obedience to the provost on their first entrance, and shall swear to observe the statutes and all and sundry the things ordained in the original foundation of the said Collegiate Church, and shall be in personal and constant residence at the said Collegiate Church by themselves, and not by others, like the rest of the prebendaries of the said church. And lest our said parish church of Dunnottir should by reason of our present erection and union be deprived of due services, and the cure of souls therein be neglected, we

debitis fraudatur obsequiis et animarum cura in ea neglegatur, statuimus et ordinamus ut dicti duo prebendarii annuatim persoluant viginti mercas monete Scotie currentis vicario pensionario pro sua sustentatione, videlicet decanus decem mercas in festo Penthecostes, et prebendarius de Donnottir alias decem mercas in festo Sancti Martini in hieme inde sequente: Quamquidem vicariam exnunc similiter erigimus et constituimus per presentes, ipseque vicarius jura nostra episcopalia archidiaconalia et alia ante presentem erectionem et vnionem consueta persoluet, curamque animarum dicte parrochialis ecclesie et parochie geret et exerceat: Collationem institutionem ac omnimodam dispositionem dictarum prebendarum ac vicarie sic per nos erectarum cum vacare contigerit, nobis et successoribus nostris, Sanctiandree archiepiscopis pro tempore existentibus, omnimodo pertinere volumus ac decernimus: Et quotienscunque contigerit nos aut aliquem ex successoribus nostris dictas prebendas aut aliquam earundem pro tempore vacantem conferre persone seu personis non qualificatis et instructis juxta tenorem antique fundationis dicte ecclesie collegiate et nostram ordinationem presentibus expressam, licebit sit preposito et prebendariis dicte ecclesie collegiate

appoint and ordain that the said two prebendaries shall pay yearly twenty merks of the current money of Scotland to a vicar pensioner for his maintenance, videlicet, the dean ten merks at the feast of Whitsunday, and the prebendary of Dunnottir other ten merks at the feast of Martinmas in the winter thereafter following; Which vicarage we now in like manner erect and constitute by these presents, and the same vicar shall pay our episcopal, archidiaconal, and other dues wont to be paid before the present erection and union, and shall bear and exercise the cure of souls in the said parish church and parish. We will and decern the collation, institution, and every kind of disposition of the said prebends and vicarage thus by us erected, when they shall happen to be vacant, to belong to us and our successors the Archbishops of St Andrews for the time being: And whenever we or any of our successors shall confer the said prebends or any of them, vacant for the time, on a person or persons not qualified and learned according to the tenor of the old foundation of the said Collegiate Church, and our ordinance expressed in these presents, it shall be lawful for the provost and prebendaries of the said Collegiate Church for the time being to decline and

pro tempore existentibus talem seu tales quibus contigerit dictas prebendas aut earum aliquam, per nos aut nostrum successorem conferri tanquam indignos et inydoneos repellere et refutare: Collationesque tales sit per nos vel successorem nostrum factas et fiendas, exnunc prout extunc et econuerso, nullas et inualidas esse decernimus: Saluis nobis et successoribus nostris juribus episcopalibus ac archidiaconalibus vel aliis de jure vel consuetudine ante presentem nostram erectionem vnionem et incorporationem debitis et consuetis. In quorum omnium et singulorum premissorum fidem ac testimonium has presentes literas siue presens publicum instrumentum erectionem et vnionem in se continentem, per notarium publicam subscriptum scribamque nostrum subscribi et publicari mandauimus, sigilliquo nostri auctentici, vnacum sigillo capituli nostri, jussimus et fecimus appensione communiri, apud Edinburgh decimoquarto die mensis Nouembris, anno Domini millesimo quingentesimo secundo, indictione sexta, pontificatus sanctissimi in Christo patris et domini nostri, domini Alexandri, diuina prouidencia Pape sexti anno vndecimo: Presentibus ibidem honorabili viro Johanne Maluill de Raith, venerabilibus et religiosis viris dompnis Roberto Suentoun sacrista de Dunfermling, Willelmo Baxtar, et Johanne Spend-

refuse as unworthy and unfit such person or persons on whom may have been conferred the said prebends or any one of them by us or our successor; and such collations made or which may be made either by us or our successors, now as then and then as now, we decern to be void and null. Saving to us and our successors the episcopal and archidiaconal rights or others due and wont by law or custom before our present erection, union, and incorporation. And in faith and testimony of all and singular the premises, this present letter or present public instrument containing in itself the erection and union, we have commanded to be subscribed and published, by the notary public subscribing, and our clerk, and we have ordered and caused it to be confirmed by the appending of our authentic seal, together with the seal of our chapter, at Edinburgh, on the fourteenth day of the month of November, in the year of our Lord one thousand five hundred and two, in the sixth indiction, and in the eleventh year of the pontificate of the most holy father in Christ and our lord the lord Alexander the sixth, by divine providence Pope. Present there an honourable man John Maluill of Raith, venerable and religious men, Sirs Robert Suentoun sacristan of Dunfermling,

luff, monachis dicti monasterii, testibus ad premissa vocatis pariter et rogatis.

Sequitur tenor subscriptionis dicti notarii in premissis.

Et ego Patricius Middiltoun in artibus magister presbyter Sanctiandree diocesis publicus apostolica et imperiali autoritatibus notarius, etc.

In quorum omnium et singulorum fidem et testimonium premissorum has presentes literas, siue presens publicum instrumentum, exinde fieri, et per notarium publicum subscriptum subscribi et publicari mandauimus, sigillique nostri officii officialatus Sanctiandree principalis quo vtimur, jussimus et fecimus appensione communiri. Datum et actum in ecclesie parrochiali Sancte Trinitatis infra ciuitatem Sanctiandree, nobis inibi pro tribunali sedentibus, sub anno a natiuitate Domini millesimo quingentesimo vigesimo quinto, mensis vero Martii die vigesimo primo, indictione decima quarta, pontificatus sanctissimi in Cristi patris et domini nostri domini Clementis, diuina prouidencia Pape septimi, anno tertio: Presentibus ibidem venerabilibus et egregiis viris Magistris Thoma Kincrage, Thoma Wemis, Georgeo Strang, Alexandro Makcahoyn, Andrea Foular, Dominis Henrico Balfour et Georgeo Gerwes

William Baxtar and John Spendluff, monks of the said monastery, witnesses to the premises, alike called and required.

Follows the tenor of the subscription of the said notary in the premises:—

And I Patrick Middletoun, master of arts, presbyter of the diocese of St Andrews, notary public, by apostolic and imperial authority, etc.

In faith and testimony of all and sundry the premises we have ordered this present letter or present public instrument to be made thereupon, and to be subscribed and published by the notary public underwritten, and have commanded and caused it to be confirmed by appending the seal which we use of our office of Official Principal of St Andrews. Given and done in the parish church of the Holy Trinity within the city of St Andrews, we sitting in judgment therein, in the year from the nativity of our Lord one thousand five hundred and twenty-five, on the twenty-first day of the month of March, in the fourteenth indiction, and in the third year of the pontificate of the most holy father in Christ and our lord the lord Clement the seventh, by divine providence Pope. Present there venerable and worthy men Masters Thomas Kincrage, Thomas Wemis, George Strang, Alexander Makcahoyn, Andrew Foular, Sirs Henry Balfour and George Gerwes,

capellanis, clericis Sanctiandree diocesis, cum diuersis aliis testibus ad premissa vocatis pariterque rogatis.

Et ego Robertus Lausone, Artium Magister, clericus Sanctiandree diocesis publicus auctoritate apostolica notarius, etc.

chaplains, clerks of the diocese of St Andrews, with divers other witnesses to the premises alike called and required.

And I Robert Lausone, Master of Arts, clerk of the diocese of St Andrews, by apostolic authority notary public, etc.

VII.

LETTER by King James V. to Pope Clement VII., praying his Holiness to grant Indulgences to those who should visit the Trinity College, and aid in the completion of the Building. Stirling, 22d March 1531.

Sanctissimo Domino nostro Pape.

BEATISSIME pater, post humillimam prostrationem ad pedes Sanctissimos. Est Collegium non incelebre juxta oppidum nostrum Edinburgi, ab olim Serenissima principe Maria Regina Scotie, proavia nostra, fundatum, cuius prepositus est Iohannes Dingvalli, vir prudens et probus, Romane Sedis prothonotarius, qui jam secum destinavit choro

To our most holy Lord, the Pope.

MOST blessed Father—after the most humble prostration at your most holy feet—there is a somewhat famous College, near to our town of Edinburgh, founded by the most serene Princess, Mary, formerly Queen of Scotland, our great grandmother, the Provost of which is John Dingwall, a prudent and good man, prothonotary of the Roman see, who has purposed with himself to make

magnifice extructo reliquum templi equare, unde opus non sine ingenti impensa absolvendum aggressus est, quod ut facilius perficiat, et ut communi juvetur auxilio, cupit ut Vestrae Sanctitatis beneficio gratiose indulgentie omnibus vere penitentibus, contritis et confessis Collegium hoc in festo dive Trinitatis, et per octavas, devotionis causa visitantibus, ac fabrice eiusdem adiutrices manus porrigentibus, pro ipsius prepositi vita tantum concedantur, possitque, per se vel idoneos presbyteros, ejusmodi indulgentias consequendi causa concurrentium confessiones audire, atque illos absolvere. Quare Vestram Sanctitatem rogamus, (ut) his tam piis desideriis benigne annuas, et indulgencias quam amplissimas, ceteraque omnia concedas, que prepositus, ipse supplicando juste postulaverit. Beatissime pater, eandem Vestram Sanctitatem Deus Optimus Maximus quam longissime felicem conservet. Ex arce nostra Stirlingensi, vigesimo secundo die mensis Martii, anno ab incarnatione Dominica trigesimo primo supra millesimum et quingentesimum.

E. V. S.

Devotus filius Scotorum Rex,
JAMES R.

the rest of the church conformable to the magnificently constructed choir, having thus undertaken a work not to be completed without great expense, which that he may the more easily accomplish, and that he may be assisted by the help of the public, he desires that by the beneficence of your Holiness, gracious indulgences may be granted, for the lifetime only of the Provost himself, to all who shall visit the College on this feast of the Holy Trinity, and during the Octaves, for the purpose of devotion, being truly penitent, contrite, and making confession, and who shall put forth helping hands to the building of the same, and that he may be empowered to hear the confessions of those who shall assemble for the sake of obtaining such indulgences, and to absolve them, either personally, or by suitable presbyters : Wherefore we crave your Holiness kindly to assent to these so pious wishes, and to grant indulgences as plenary as possible, and all other things, which the Provost himself may with justice humbly crave. Most blessed Father, may God Almighty very long preserve your Holiness in felicity. From our castle at Stirling on the 22d day of the month of March in the year One thousand five hundred and thirty-one from our Lord's incarnation.

Your most excellent Holiness'

Devoted Son, the King of Scots,
JAMES R.

VIII.

Charter by Queen Mary granting the Kirk-livings to the Provost, Bailies, Council, and Community of the Burgh of Edinburgh. Edinburgh, 13th March 1566-7.

Maria Dei gratia Regina Scotorum: Omnibus probis hominibus totius terre sue clericis et laicis, salutem. Sciatis quia nos impensius munus nostrum erga divinum servitium perpendentes, et pro ardenti zelo quem ob intertenendam policiam et equabilem ordinem inter subditos nostros, precipue vero infra Burgum nostrum de Edinburgh, preservandum habemus; considerantes itaque nos ex officio teneri (et) munus erga Deum complecti debere, cujus providentia regimini hujus regni proponimur, sicque nobis ex officio incumbere omni honesto modo pro ministris verbi Dei providere, et quod hospitalia pauperibus mutilatis et miseris personis, orphanis et parentibus destitutis infantibus, infra dictum nostrum burgum preserventur, post nostram perfectam etatem, cum avisamento dominorum secreti consilii nostri, dedimus concessimus disposuimus, ac pro nobis et successoribus nostris pro perpetuo confirmavimus, necnon

Mary by the grace of God Queen of Scots: To all good men of her whole land, clerics and laics, greeting. Know ye that we more carefully reflecting upon our duty towards the service of God, and out of the ardent zeal which we have for maintaining the civil polity, and preserving good order among our subjects, but chiefly within our Burgh of Edinburgh, and also considering that we by our office are bound and ought to be careful of our duty towards God, by whose providence we are set over the government of this kingdom, and that it is incumbent on us in virtue of our office, by all honest means to provide for the ministers of God's word, and that hospitals for poor mutilated and miserable persons, orphans and children deprived of their parents, may be maintained within our said Burgh, did, on attaining our majority, with the advice of the lords of our Privy Council, give, grant, dispone, and for us and our successors for ever

tenore presentium damus concedimus disponimus, et pro nobis et nostris successoribus pro perpetuo confirmamus predilectis nostris Preposito Ballivis Consulibus et Communitati dicti nostri Burgi de Edinburgh et ipsorum successoribus imperpetuum, Omnes et Singulas terras tenementa domos edificia ecclesias capellas hortos pomaria croftas annuos redditus fructus devoria proficua emolumenta firmas elimozinas lie daill silver obitus et anniversaria quecunque, que quovismodo pertinuerunt aut pertinere dinoscuntur ad quascunque capellanias alteragia prebendarias, in quacunque ecclesia capella aut collegio infra libertatem dicti nostri Burgi de Edinburgh fundata seu fundatas per quemcunque patronum, in quarum possessione capellani et prebendarii earundem perprius fuerant, ubicunque prefate domus tenementa edificia pomeria horti annui redditus anniversaria fructus proventus et emolumenta jacent aut prius leuata fuerunt respective, cum maneriis locis pomeriis terris annuis redditibus emolumentis et devoriis quibuscunque que Fratribus Dominicalibus seu Predicatoribus et Minoribus seu Franciscanis dicti nostri Burgi de Edinburgh perprius pertinuerunt; unacum omnibus et singulis terris domibus tenementisque jacentibus infra dictum nostrum Burgum et

confirm, and do by the tenor of these presents give, grant, dispone, and for us and our successors for ever confirm to our well beloved the Provost, Bailies, Councillors, and Community of our said Burgh of Edinburgh, and their successors for ever, All and Singular the lands, tenements, houses, buildings, churches, chapels, yards, orchards, crofts, annual rents, fruits, duties, profits, emoluments, rents, alms, daill-silver, obits, and anniversaries whatsoever, which anywise belonged or are known to belong to any chaplainries, altarages, and prebends, founded in any church, chapel, or college within the liberty of our said Burgh by whatsoever patron, in possession whereof the chaplains and prebendaries of the same formerly were, wheresoever the foresaid houses, tenements, buildings, orchards, yards, annual rents, anniversaries, fruits, profits, and emoluments lie, or were formerly uplifted respectively, with the manor places, orchards, lands, annual rents, emoluments, and duties whatsoever which formerly belonged to the Dominican or Preaching Friars, and to the Minorites or Franciscans of our said Burgh of Edinburgh; together with all and sundry lands, houses, and tenements lying within our said Burgh and the liberty of the same, with all annual rents leviable from any house, lands, or tenement

libertatem ejusdem, cum omnibus annuis redditibus de quacunque domo terris aut tenemento infra dictum nostrum Burgum leuandis, datis fundatis et donatis quibuscunque capellaniis alteragiis ecclesiis mortuariis aut anniversariis, ubicunque sint infra regnum nostrum; ac etiam cum omnibus et singulis annuis redditibus et aliis devoriis solitis aut que per quamcunque ecclesiam extra dictum nostrum Burgum a Preposito aut Ballivis ejusdem de communi redditu ejusdem pro suffragiis celebrandis demandari poterint, cum pertinentiis. TENENDAS et HABENDAS omnes et singulas prefatas terras tenementa domos edificia pomeria hortos croftas annuos redditus fructus devoria proficua emolumenta firmas elemozinas obitus anniversaria ecclesias capellas fratrum loca hortos cum pertinentiis prefatis Preposito Ballivis Consulibus et Communitati et eorum successoribus de nobis et successoribus nostris imperpetuum, prout eadem jacent in longitudine et latitudine, in domibus edificiis muris muremiis lignis lapide et calce, cum libero introitu et exitu etc. ac cum omnibus aliis et singulis libertatibus commoditatibus proficuis et asiamentis ac justis suis pertinentiis quibuscunque, tam non nominatis quam nominatis, tam sub terra quam supra terram, ad predictas terras tenementa domos edificia pomeria hortos croftas annuos redditus fructus

within our said Burgh, given, founded, and granted to whatever chaplainries, altarages, churches, burials or anniversaries, wheresoever they may be within our kingdom, and also with all and sundry annual rents and other dues customary, or that could be demanded by any church outwith our said Burgh from the Provost or bailies of the same out of the common good of the same for celebrating suffrages, with the pertinents. To HOLD and TO HAVE all and singular the foresaid lands, tenements, houses, buildings, orchards, yards, crofts, annual rents, fruits, duties, profits, emoluments, rents, alms, obits, anniversaries, churches, chapels, friars' places, and yards with the foresaid pertinents to the Provost, Bailies, Councillors, and Community, and their successors, of us and our successors for ever, as the same lie in length and breadth in houses, buildings, walls, timber, wood, stone, and lime, with free ish and entry, etc., and with all and sundry liberties, commodities, profits, and easements, and their just pertinents whatsoever, as well not named as named, as well under the ground as above the ground, belonging to the foresaid lands, tenements, houses, buildings, orchards, yards, crofts, annual rents, fruits, duties, and other things aforesaid,

devoria et alia prescripta cum pertinentiis spectantibus seu juste spectare valentibus quomodolibet, in futurum, libere quiete plenarie integre honorifice bene et in pace absque revocatione aut contradictione quacunque. Cum potestate memoratis Preposito Ballivis Consulibus et communitati et ipsorum successoribus, per seipsos et ipsorum collectores quos constituent, prefatos annuos redditus fructus devoria proficua emolumenta quecunque levandi et recipiendi ubicunque perprius levata fuerant, prefatas terras et tenementa locandi et removendi, loca diruta extruendi et reparandi, eademque in hospitalia aut alios similes usus legitimos, prout ipsis cum avisamento ministrorum et seniorum dicti nostri Burgi videbitur, reducendi et applicandi, adeo libere in omnibus sicuti prefati prebendarii capellani et fratres prescripti eisdem perprius gaudere easdemque possidere potuissent; memorati autem Prepositus Ballivi Consules et eorum successores tenebuntur et astricti erunt ministros lectores et alia ecclesiastica onera prefatis annuis redditibus proficuis et devoriis secundum valorem et quantitatem earundem sustinere, locaque et edificia reparanda in hospitalitatem et alios usus prescriptos applicare. Considerantes itaque quanta fraude ingens

with their pertinents, or which may justly belong thereto in any manner of way, freely, quietly, fully, wholly, honourably, well, and in peace, for the time to come, without revocation or challenge whatsoever. With power to the above mentioned Provost, Bailies, Councillors and Community, and their successors, by themselves and their collectors whom they shall appoint, to uplift and receive the said annual rents, fruits, duties, profits, and emoluments whatsoever, wherever they were formerly uplifted, to let and remove [from] the foresaid lands and tenements, to build and repair the ruinous places, and to restore and apply the same to hospitality or other similar lawful uses, as to them, with the advice of the ministers and elders of our said Burgh, shall seem fit, as freely in all respects as the said prebendaries, chaplains, and friars before written might have enjoyed and possessed the same aforetime. Moreover, the said Provost, Bailies, Councillors, and their successors shall be holden and obliged to support the ministers, readers, and other ecclesiastical charges out of the said annual rents, profits, and duties, according to the value and quantity of the same, and to apply the places and buildings to be repaired for hospitality and other uses foresaid. Besides, considering how dishonestly a great number of the said prebendaries, chap-

numerus dictorum prebendariorum capellanorum et fratrum prescriptorum, qui post alterationem religionis terras annuos redditus et emolumenta ipsorum capellaniis prebendis et aliis locis respective perprius mortificata disposuerunt alienarunt et in manibus quorundam particularium virorum extradonarunt ; ac etiam quod plerique legii nostri quarundem terrarum tenementorum et annuorum redditunm per ipsorum predecessores mortificatorum jus sibi acclamarunt per brevia capelle nostre aut alias sasinam tanquam heredes suorum predecessorum (qui easdem ecclesie perprius dotarunt) recuperarunt, quod evenit partim negligentia officiariorum dicti nostri Burgi et partim collusione dictorum prebendariorum capellanorum et fratrum prescriptorum. Quocirca cum avisamento prescripto omnes et singulas hujusmodi alienationes dispositiones et sasinas quibus primum propositum et animus fundatorum infringitur alteratur et variatur, deducendo easdem in particulares usus, ad effectum quod eedem in usus suprascriptos converti poterint per presentes rescindimus et annullamus. Quamquidem hanc nostram declarationem volumus tanti esse roboris et efficacie ac si persone que easdem dispositiones obtinuerunt particulariter citate essent ipsarumque infeofamenta absque ulteriori processu rescinderentur.

lains, and friars foresaid, have, since the change of religion, disponed, alienated, and given away into the hands of certain particular persons, the lands, annual rents, and emoluments previously mortified to their chaplainries, prebends, and other places respectively ; and also that very many of our lieges have claimed for themselves, by brieves of our chancery, the right to certain lands, tenements, and annual rents mortified by their predecessors, or otherwise have obtained sasine as heirs of their predecessors, who previously gifted the same to the church, which has happened partly by the negligence of the officers of our said Burgh and partly by the collusion of the said prebendaries, chaplains, and friars, foresaid. Wherefore, with advice aforesaid, we, by these presents, rescind and annul all and sundry such alienations, dispositions, and sasines, by which the first purpose and will of the founders is infringed, altered, and changed, by perverting the same to individual (or private) uses, to the effect that the same may be converted to the purposes above set forth. And this our declaration we will to be as strong and effectual as if the persons who obtained the same dispositions had been particularly cited and their infeftments rescinded without further process. As also, with

Ac etiam cum avisamento prescripto unimus et incorporamus omnes et singulas terras tenementa domos edificia ecclesias cimiteria capellas pomaria hortos croftas annuos redditus fructus devoria proficua emolumenta firmas elemozinas obitus anniversaria fratrum loca hortos eorundem cum suis pertinentiis in unum corpus, imposterum appellandum Fundatio nostra Ministerii et Hospitalitatis de Edinburgh. Volumus etiam quod unica sasina, per prefatos Prepositum et Ballivos aut ipsorum aliquem, dicti ministerii et hospitalitatis nomine, apud Pretorium dicti nostri burgi semel accepta, tam sufficiens erit sasina perpetuo in futurum ac si eadem super particulares terras ad dictos capellanos prebendarios [et] fratres pertinentes aut ipsis in prefatos annuos redditus anniversaria firmas proficua et devoria prescripta debitas sumeretur, non obstante locorum distantia. Preterea per presentes nolumus capellanos prebendarios et fratres qui ante dictam alterationem provisi erant per hoc presens nostrum infeofamentum prejudicari, sed reservamus illis usum dictorum fructuum et devoriorum durante eorum vita tantum. PRECIPIENDO itaque nostrorum computorum rotulatoribus presentibus et futuris ipsorum collectoribus factoribus et aliis quorum

advice foresaid, we unite and incorporate all and singular, the lands, tenements, houses, buildings, churches, cemeteries, chapels, orchards, yards, crofts, annual rents, fruits, duties, profits, emoluments, rents, alms, obits, anniversaries, friars' places, yards of the same, with their pertinents, into one body to be called in all time coming our Foundation of the Ministry and Hospitality of Edinburgh. We will also that one sasine, taken once for all at the Tolbooth of our said burgh by the foresaid Provost and Bailies, or any of them, in name of the said ministry and hospitality, shall be as sufficient sasine for all time coming as if the same were taken upon the particular lands belonging to the said chaplains, prebendaries, and friars, or in the foresaid annual rents, anniversaries, rents, profits, and duties foresaid due to them, the distance of the places notwithstanding. Besides, by these presents, we will that no injury be done to the chaplains, prebendaries, and friars who were in possession before the said change of religion by this our present infeftment, but we reserve to them the use of the foresaid fruits and duties during their lives only. Directing, accordingly, our comptrollers, present and future, and their collectors, factors, and others

interest in genere necnon in specie, quod ne quis eorum recipere aut levare presumat dictos fructus particulariter suprascriptos pro quovis tempore preterito seu futuro, neve impediant aut impedimentum ullum faciant memoratis Preposito Ballivis Consulibus Communitati et ipsorum successoribus in pacifica possessione eorundem. Requirendo et ordinando etiam dominos Sessionis nostre quatenus literas in omnibus quatuor formis ad instantiam dictorum Prepositi Ballivorum Consulum Communitatis et ipsorum successorum ad effectum suprascriptum dirigant. Necnon precipiendo quibuscunque intromissoribus cum dictis fructibus quatenus ipsis de eisdem prompte intendant pareant et gratam solutionem faciant. In cujus rei testimonium huic presenti carte nostre magnum sigillum nostrum apponi precepimus. Testibus, reverendissimo in Christo patre Johanne archiepiscopo Sanctiandree etc.; dilectis nostris consanguineis, Georgio comite de Huntlie domino Gordoun et Badyenacht, cancellario nostro, Jacobo comite de Boithuill domino Haillis Creychtoun et Liddisdaill, regni nostri magno admirallo; dilectis nostris familiaribus consiliariis, Richardo Maitland de Lethingtoun, nostri secreti sigilli custode, Jacobo Balfoure de Pettindreich, nostrorum rotulorum registri ac consilii clerico, et Johanne Bellenden de Auchnoule

whom it concerns in general as well as in special, that none of them presume to receive or to levy the said fruits particularly above described for any time whatsoever, past or future, or offer any obstruction or impediment to the foresaid Provost, Bailies, Counsellors, Community and their successors in the peaceable possession of the same. Requiring and ordaining also our lords of Session that they direct letters in all the four forms at the instance of the said Provost, Bailies, Councillors, Community and their successors to the effect above written. Also commanding all intromitters with the said fruits that that they give prompt attention, (and that they) obey, and make willing and ready payment to them of the same: In witness whereof we have commanded our great seal to be affixed to this our present charter. Witnesses, the most reverend father in Christ John archbishop of St Andrews, our beloved cousins, George earl of Huntly, lord Gordon and Badyenacht, our chancellor, James earl of Bothwell, lord Hailes, Crichton and Liddisdale, high admiral of our kingdom, our familiar councillors, Richard Maitland of Lethington, keeper of our privy seal, James Balfour of Pittendreich, clerk of our rolls register and council, and John

nostre justiciare clerico, equitibus auratis; Apud Edinburgh decimo tertio die mensis Martii anno Domini millesimo quingentesimo sexagesimo sexto et regni nostri vicesimo quinto.

Bellenden of Auchnoull, our justice-clerk, knights. At Edinburgh, the thirteenth day of the month of March in the year of our Lord One thousand five hundred and sixty-six, and the twenty-fifth of our reign.

IX.

LETTERS of Remission by King James VI., dispensing with the erection of an Hospital on the Blackfriars Yards, and granting the site and buildings, &c., of Trinity College to the Provost, Bailies, Council and Community of Edinburgh for the purposes of an Hospital. 3d January 1566–7.

JACOBUS Dei gratia Rex Scotorum: Universis et singulis ad quorum notitias presentes litere pervenerint, salutem. Universitati nostre constet et notum sit quod charissima nostra mater pro tempore Regina per suam cartam suo sub magno sigillo Totum et Integrum locum et pomeria que fratribus the Blackfreiris, predicatoribus olim nuncupatis, burgensibus Burgi nostri de Edinburgh pertinuerunt, cum cimiteriis et aliis pertinentibus ejusdem, pro constructione et plantatione unius Hospitalis

JAMES, by the grace of God King of Scots: To all and suudry to whose knowledge the present letter shall come, greeting. Be it certified and known to our whole community, that our dearest mother, for the time Queen, did, by her charter, under her great seal, give, grant, and dispone All and Whole the place and yards which belonged to the Black friars, formerly called Friars Preachers, burgesses of our Burgh of Edinburgh, with the cemeteries and other pertinents of the same, for the construction and erection of an Hospital

desuper pro adjuvamine seu auxilio pauperum, Preposito Ballivis Consulibus et Communitati dicti Burgi nostri de Edinburgh presentibus et eorum successoribus qui pro tempore fore contigerint, de nobis et successoribus nostris in libero burgagio pro servitio burgi usitato et consueto tenenda hereditarie dedit, concessit, et disposuit. Et quod in dicto infeofamento promissum est quod ipsi dictum Hospitale infra annum unum post datam dicti infeofamenti inchoarent ac idem intra spatium decem annorum depost proxime sequentium complerent, quemadmodum in memorato infeofamento sub magno sigillo de data apud Sanctumandream decimo sexto die mensis Martii anno Domini millesimo quingentesimo sexagesimo secundo latius continetur. Et quia nobis et Regenti antedicto evidenter monstratum et declaratum extat quod prefatus locus fratrum non tam commodus nec conveniens pro edificatione vel constructione et erectione dicti Hospitalis sicuti locus Collegii divini Trinitatis, orti, domus, et edificia ejusdem existit, ratione grandis frequentationis et confluentiæ tam alienigenorum quam legeorum nostri Regni qui prope per eundem indies accedunt, vel pertranseunt et frequentant, per quem seu quod emolumentum et auxilium ad dictum

thereupon, for the relief or assistance of the poor, to the present Provost, Bailies, Councillors and Community of our said Burgh of Edinburgh, and those who shall happen to be their successors for the time being, to be held perpetually of us and our successors in free burgage, for service of burgh used and wont. And because it was promised in the said infeftment that they should begin the said Hospital within one year after the date of the said infeftment, and should finish the same within the space of the ten years thereafter next following, as in the said infeftment, under the great seal, dated at St Andrews, on the sixteenth day of the month of March, in the year of our Lord One thousand five hundred and sixty-two, is more fully set forth; and because it is clearly shewn and declared to us and the aforesaid Regent, that the friars' place before mentioned is not so suitable or convenient for the building or construction and erection of the said Hospital as the place of the College of the Holy Trinity, and the yards, houses, and buildings of the same, by reason of the great crowd and concourse, both of strangers and of the lieges of our kingdom, who daily approach or pass through and frequent the same, by whom some profit and aid may daily accrue to the said Hospital and

hospitale quotidie redundare et indigentibus personis in eodem pervenire poterit. Quodquidem dicti Prepositus Ballivi et Consules et Communitas in promptu construere et reparare inchoarunt, atque volentes inde quod proficuum et commoditas illius unius loci seu bondarum aut continentie cum loco fratrum super dicto Hospitali Collegii prescripti expendantur et exponantur. IDEO, et pro diversis aliis bonis seu divinis occasionibus et considerationibus nos et memoratum Regentem nostrum moventibus, cum prelibatis Preposito Ballivis Consulibus et Communitate dicti Burgi nostri de Edinburgh et eorum successoribus, penes illam partem et clausulam predicti infeofamenti inchoationem completionem ac edificationem antedicti Hospitalis concernentem super loco et bondis prefatorum fratrum intra spatium prescriptum, atque penes omne damnum lesionem et prejudicium que ipsis eorum infeofamento supradicto pro non completione seu perimpletione ejusdem succedere et sequi poterunt, tenore presentium dispensavimus ac dispensamus. NECNON volumus et concedimus per presentes quod dictum infeofamentum in robore et effectu iis et eorum successoribus stet et remaneat non obstante dicta clausula ac provisione in eodem, ut

reach the indigent persons within the same; which [Hospital] the said Provost, Bailies, Councillors and Community have promptly commenced to build and repair, and the more willingly, because the profit and advantage of that particular site, or of its bounds or continuity with the friars' place, may be expended and laid out on the said Hospital of the aforementioned College: Therefore, and for divers other good or pious reasons and considerations moving us and our Regent aforesaid, by the tenor of these presents, we have dispensed, and do dispense, in favour of the said Provost, Bailies, Councillors and Community of our said Burgh of Edinburgh, with that part and clause of the foresaid infeftment relating to the commencement, completion, and building of the foresaid Hospital upon the place and bounds of the forementioned friars within the time before-written, and with all loss, injury, and prejudice which can accrue to and fall on them from their infeftment above mentioned, for the non-completion or implementing of the same. Moreover, we, by these presents, will and grant that the said infeftment shall stand and abide in force and effect to them and their successors, the said clause and provision contained in the same, as aforesaid, notwithstanding, with which, by these presents, we for ever

prefertur, contenta, cum hujusmodi (clausula) per presentes perpetuo dispensamus; ac etiam damus, concedimus, et committimus antedicto Preposito, et Ballivis, Consulibus, et Communitati, eorumque successoribus, nostram plenam licentiam et potestatem super prefato loco pomerio ortis, domibus, cemeterio et hujusmodi pertinentibus ad eorum majorem commoditatem et proficuum et valorem, in feodifirma annuo redditu annuatim assedandi utendi et quemadmodum eiis expediri videbitur disponendi, sic quod annuum proficuum redditus et devoria eorundem ad sustentationem dicti hospitalis divini Trinitatis Collegii et pauperum in eodem existentium ac nullis aliis redundare et applicari poterint; Ratificamus approbamus confirmamus nunc prout ex tunc, et ex tunc prout ex nunc, omnia talia feuda sive feodifirmas et dispositiones per ipsos Prepositum Ballivos Consules Communitatem et eorum successores, de prefatorum loco fratrum ortis domibus cemeterio et pertinentibus, quibuscunque persone vel personis, ad effectum supra specificatum duntaxat factas seu faciendas. Datum sub testimonio nostri magni sigilli apud Edinburgh tertio die mensis Januarij anno Domini millesimo quinquagesimo sexagesimo sexto [septimo ?] et regni nostri primo.

dispense; and we also, give, grant, and commit to the aforesaid Provost and Bailies, Councillors and Community, and their successors, our full license and power of leasing, using, and disponing upon the foresaid place, yard, gardens, houses, cemetery, and such pertinents, in feu-farm [or] for a yearly rent, as shall seem to them expedient, for their greater advantage and profit, so that the annual profit, rents, and duties of the same shall increase and be applied to the sustentation of the said Hospital of the College of the Holy Trinity, and of the poor residing in the same, and of no others. And we then as now, and now as then, ratify, approve, and confirm all such feus or feu-farms and dispositions of the place of the foresaid friars, gardens, dwellings, cemetery, and pertinents, made or to be made, to the effect above specified only, by the said Provost, Bailies, Councillors, Community, and their successors, to whatsoever person or persons. Given under the testimony of our great seal, at Edinburgh, the third day of the month of January, in the year of our Lord One thousand five hundred and sixty-six [seven ?], and the first of our reign.

X.

Charter by King James VI. granting Trinity Church and Hospital to Sir Simon Preston, provost, and his successors the Provosts, Bailies, and Council of the Burgh of Edinburgh. Edinburgh, 12th November 1567.

JACOBUS Dei gratia Rex Scotorum: Omnibus probis hominibus totius terre sue clericis et laicis, salutem. Sciatis quod nos et charissimus consanguineus Jacobus comes Morauie dominus Abirnethy etc. nostri regni Regens animo ferventi et zelo ducti ad supportandum et adjuvandum paupertatem penuriam et inopiam multarum et diversarum honestarum senium et impotentium personarum a quibus in earum senectute per eventum et adversam fortunam res et bona deciderunt, ne propter extremam famem penuriam et indigentiam sue necessarie sustentationis omnino perirent et morirentur, Nos propterea pietate et bona conscientia moti ad prestandum eis juvamen et auxilium prout eorum indigentia et necessitas requirit, ac etiam intelligentes quod hec predicta in omnibus commoda principia et initia capere non poterunt nec commode performari et ad finem perfectum pervenire valeant absque

JAMES, by the grace of God King of Scots: To all the good men of his whole land, clerics and laics, greeting. Know ye that we and our dearest cousin James earl of Murray lord Abernethie, &c., Regent of our kingdom, moved by fervent and zealous purpose to support and assist the poverty, penury, and want of many and divers honest, aged, and impotent persons, who in their old age have lost their means and substance by accident and bad fortune, so that they may not utterly perish and die through extreme hunger, penury, and want of their necessary sustenance; we therefore, moved by piety and good conscience to afford them such help and assistance as their indigence and necessity requires; as also understanding that the aforesaid purpose cannot in all respects be conveniently begun and commenced, nor conveniently perfected and accomplished, without our

nostro supplemento auxilio et auctoritate; intelligentesque quod dominus Symon Prestoun de eodem miles animo est deliberato firmo et constanti proposito ad edificandum construendum ac cum omni cura et diligentia dotandum unum Hospitale cum rationabili sustentatione talibus predictis honestis indigentibus et impotentibus personis senibus et etate provectis egrotis, incolis et inhabitantibus infra nostrum Burgum de Edinburgh, ac etiam aliis senibus indigentibus et impotentibus qui idonei fuerint inventi ad acceptandum talia beneficia et gratitudines in dicto hospitali fundando. Nos propterea et Regens noster predictus intelligentes predictum propositum et opus omnibus modis non solum bonum et divinum fore, sed etiam volentes prestare occasionem aliis nostris subditis et ad alliciendos animos quorundam aliorum nostrorum ligeorum et subditorum ad simile propositum et opus talem divinam vocationem acceptandum, cum avisamento et consensu dominorum nostri secreti consilii expediens et necessarium fore duximus ad gratificandum dictum dominum Symonem prepositum dicti nostri Burgi de Edinburgh dono et donatione talis loci nunc in nostris manibus vacantis et ad nostram donationem spectantis et pertinentis, magis convenientis et idonei ad construendum et edificandum reparandum et

supplement, aid, and authority; and understanding that Sir Simon Prestoun of that Ilk, knight, intends with deliberate, firm, and set purpose to build, erect, and with all care and diligence endow an Hospital, with reasonable support for such foresaid honest poor and impotent persons, aged and advanced in years, or sick, indwellers and inhabitants within our Burgh of Edinburgh, and also for such other old indigent and impotent people as shall be found fit for receiving such benefits and charity in the said Hospital so to be founded: Therefore, we and our foresaid Regent perceiving that the said purpose and work will be in every respect not only good and divine, but also willing to give occasion to others our subjects, and to incline the minds of certain others of our lieges and subjects to accept such a divine call to a similar purpose and work, with the advice and consent of the lords of our Privy Council, have deemed it expedient and necessary to gratify the said Sir Simon, provost of our said Burgh of Edinburgh, with the gift and donation of such a place now vacant in our hands, and belonging and pertaining to our gift, as shall be most fit and convenient for constructing and building, repairing and perfecting the said Hospital, with the houses,

performandum dictum Hospitale cum domibus edificiis et hortis eiusdem, ubi major populi et gentium multitudo et confluentia tam extraneorum quam aliorum nostrorum ligeorum huius nostri oppidi frequentare videntur, et prope dictum locum quotidianum accessum ad dictum nostrum oppidum ac etiam regressum a dicto nostro oppido maxime habent, occasione cuius quotidiana elemosyna et auxilium ad dictum Hospitale augetur et increscet. Quare nos propter bonum fidele et gratuitum servitium nobis nostro Regenti predicto et etiam predictis dominis nostri secreti consilii per dictum dominum Symonem Prestoun prepositum predictum temporibus retroactis et preteritis factum et impensum, ac etiam propter nonnullas alias occasiones et considerationes animum nostrum moventes erga dictum dominum Symonem Prestoun Prepositum Ballivos Consules et Communitatem dicti nostri Burgi de Edinburgh, dedimus concessimus et disposuimus, ac tenore presentis carte nostre, damus concedimus et disponimus dicto domino Symoni Prestoun nunc preposito dicti nostri Burgi de Edinburgh et successoribus suis Prepositis Ballivis Consulibus et Communitati eiusdem Burgi pro tempore existentibus, totam et integram ecclesiam vocatam Ecclesiam Collegiatam Trinitatis cum cimiterio domibus edificiis ruinatis et

buildings and yards thereof, which the greatest multitude and concourse of people, as well strangers as others our lieges of this our town, are seen to frequent, and near which they chiefly have daily access to and egress from the town, whereby the daily alms and contributions to the said Hospital are increased and will increase: Therefore, for the good, faithful, and gratuitous service rendered and performed by the said Sir Simon Prestoun, provost foresaid, towards ourself, our foresaid Regent, and the said lords of our Privy Council, in times bygone and past, as well as for other causes and considerations moving us in favour of the said Sir Simon Prestoun, Provost, the Bailies, Councillors, and Community of our said Burgh of Edinburgh, we have given, granted, and disponed, and, by the tenor of our present charter, give, grant, and dispone, to the said Sir Simon Prestoun, present provost of our said Burgh of Edinburgh, and his successors, the Provosts, Bailies, Councillors, and Community of the said Burgh for the time being, All and Whole the church called the Collegiate Church of the Trinity, with the churchyard, houses, buildings, ruinous and built, orchards, yards, crofts, dovecot, and pertinents thereof

edificatis pomariis hortis croftis columbario et pertinentiis eiusdem quibuscunque per prepositum et prebendarios dicte Ecclesie Collegiate perprius occupatis et inhabitatis, unacum loco et parte cum edificiis et hortis hospitalis vocati Hospitalis Trinitatis vulgo Trinitie Hospitall dicte Ecclesie Collegiate contigue adjacentibus cum horto ex parte occidentali eiusdem jacente ad caudam sive finem vici seu vinelle nostre vocati Leyth Wynde, in manibus nostris nunc existente et ad nostram donationem seu dispositionem deveniente tanquam prefati Collegii et loci indubitati patroni, per ordinem actorum et statutorum a tempore reformationis religionis nuper factorum et ordinatorum, ac pro edificatione et constructione dicti hospitalis domorum hortorum et policiorum eiusdem pro sustentatione pauperum et egrotorum per ipsos infra eandem locandorum et nulli alio usui tantummodo. Tenendam et habendam totam et integram prefatam ecclesiam vocatam Ecclesiam Trinitatis cum hortis domibus edificiis pomariis croftis columbario ac domibus dicti hospitalis vocati Hospitalis Trinitatis cum omnibus locis partibus et possessionibus earundem per prepositum et prebendarios dicte Ecclesie Collegiate perprius occupatis et possedatis [possessis], dicto domino Symoni Prestoun nunc preposito dicti nostri Burgi de Edinburgh et successoribus suis Pre-

whatsoever, formerly occupied and inhabited by the provost and prebendaries of the said Collegiate Church, together with the place and part, with the buildings and yards of the hospital, called the Hospital of the Trinity, lying contiguous to the said Collegiate Church, with the yard lying on the west side thereof, at the foot or end of our street or vennel called Leith Wynd, now in our hands, and at our gift or disposal as undoubted patron of the said College and place, according to the tenor of the acts and statutes made and ordained shortly after the time of the Reformation of religion, and for the building and construction of the said Hospital, houses, yards and policies of the same, for the sustentation of the poor and sick to be placed by them within the same only, and for no other use whatever. To have and to hold all and whole the said church called the Church of the Trinity, with the yards, houses, buildings, orchards, crofts, dovecot, and houses of the said hospital called Trinity Hospital, with all the places, parts and possessions of the same, formerly occupied and possessed by the provost and prebendaries of the said Collegiate Church, to the said Sir Simon Prestoun, now provost of our said Burgh of Edinburgh, and

positis Ballivis Consulibus et Communitati dicti nostri Burgi pro tempore existentibus, de nobis et successoribus nostris in libera alba firma imperpetuum per omnes rectas metas et divisas prout prefata ecclesia cum pomariis hortis columbario et aliis prescriptis et earundem pertinentiis jacent in longitudine et latitudine in domibus edificiis hortis etc. cum libero introitu et exitu viis et passagiis earundem usitatis et consuetis cum omnibus aliis et singulis commoditatibus libertatibus asiamentis privilegiis et justis suis pertinentiis quibuscunque spectantibus seu juste spectare valentibus seu que in futurum pertinere dinoscuntur, libere quiete plenarie integre honorifice bene et in pace absque revocatione aut contradictione quacunque; cum plenaria potestate dicto domino Symoni nunc preposito et successoribus suis Prepositis Ballivis Consulibus et Communitati dicti Burgi pro tempore existentibus desuper disponendi prout ipsis visum fuerit; proviso tamen quod astricti erunt ut loca et alia prescripta usui prescripto et nullo alio modo nec usui applicabuntur. Reddendo inde annuatim dictus dominus Symon Prestoun nunc prepositus dicti nostri Burgi de Edinburgh et successores sui Prepositi Ballivi Consules et Communitas dicti nostri Burgi pro tempore

his successors, the Provosts, Bailies, Councillors, and Community of our said Burgh of Edinburgh for the time being, of us and our successors in free blench farm for ever, by all the just marches and divisions, as the foresaid church, with the orchards, yards, dovecot, and others before writtèn, and their pertinents, lie in length and breadth, in houses, buildings, yards, etc., with free ish and entry, ways and passages of the same used and wont, with all and singular commodities, liberties, easements, privileges, and their just pertinents whatsoever, belonging, or which ought justly to belong, or which are known to belong, to the same, in future, freely, quietly, fully, wholly, honourably, well and in peace, without any revocation or gainsaying whatsoever; with full power to the said Sir Simon, now provost, and his successors, the Provosts, Bailies, Councillors and Community of the said Burgh for the time being, to dispone thereupon as to them shall seem good: providing always that they shall be bound to apply the places and others foresaid to the use before set forth, and to no other purpose. Giving therefor yearly, the said Sir Simon Prestoun, now provost of our said Burgh of Edinburgh, and his successors, the Provosts, Bailies, Councillors and Community of our said Burgh for the time being, to us and our successors, a silver penny, on the

existentes, nobis et successoribus nostris unum denarium argenti super fundo prefati loci in festo Penthecostes nomine albe firme si petatur tantum. Proviso omnimodo quod hec presens donatio et dispositio, preposito et prebendariis dicte Ecclesie Collegiate, juxta ipsorum infeofamenta jura et donationes tantorum pauperum vocatorum vulgo beidmen in dicto hospitali, vocato The Trinitie Hospitale predicto, nunc locatorum et dotatorum, secundum tenorem erectionis desuper facte, minime prejudicet. In cuius rei testimonium huic presenti carte nostre magnum sigillum nostrum apponi precepimus. Testibus, reverendissimo in Christo patre Johanne Archiepiscopo Sanctiandree etc.; dilectis nostris consanguineis Jacobo comite de Mortoun domino Dalkeyth cancellario nostro, Wilelmo comite Mariscalli domino Keyth; venerabili in Christo patre Johanne commendatario monasterii nostri de Coldinghame nostri secreti sigilli custode; dilectis nostris familiaribus consiliariis Magistro Jacobo McGill de Rankelour Nether nostrorum rotulorum registri ac consilii clerico, et Johanne Bellenden de Auchnoule milite nostre justiciarie clerico. Apud Edinburgh duodecimo die mensis Novembris anno Domini millesimo quingentesimo sexagesimo septimo, et regni nostri anno primo.

ground of the said place and others, at Whitsunday, in name of blench farm, if asked only. Providing always that this present gift and grant shall be in no degree prejudicial to the provost and prebendaries of the said Collegiate Church, in regard to their infeftments, rights, and donations to so many of the poor, commonly called beidmen, now placed and endowed in the said hospital, called the Trinity Hospital foresaid, after the tenor of the erection made thereupon. In witness whereof we have ordered our Great Seal to be appended to this our present charter. Witnesses, the most reverend father in Christ John archbishop of St Andrews, etc.; our beloved cousins, James earl of Mortoun lord Dalkeith, our chancellor, William earl Marischall lord Keith; the venerable father in Christ John commendator of our monastery of Coldingham, keeper of our privy seal; our beloved familiar councillors, Mr James M'Gill of Rankeillour Nether, clerk of our rolls register and council, and John Bellenden of Auchnoule knight, our justice clerk. At Edinburgh, the twelfth day of the month of November, in the year of our Lord one thousand five hundred and sixty-seven, and in the first year of our reign.

XI.

CHARTER by King James VI. confirming Queen Mary's charter of 13th March 1566, and of new granting the Kirk-livings to the Provost, Bailies, Council, and Community of the Burgh of Edinburgh. Stirling, 14th April 1582.

JACOBUS Dei gratia Rex Scotorum: Omnibus probis hominibus totius terre sue clericis et laicis, salutem: SCIATIS nos, cum avisamento dominorum nostri secreti consilii, quandam cartam et infeofamentum per nostram charissimam matrem pro tempore regni nostri Reginam post suam perfectam etatem cum avisamento et consensu dominorum ejus secreti consilii factam datam et concessam dilectis nostris Preposito Ballivis Consulibus et Communitati Burgi nostri de Edinburgh et eorum successoribus super donatione dispositione et confirmatione omnium et singularum terrarum tenementorum domorum edificiorum ecclesiarum capellaniarum hortorum pomariorum croftarum annuorum reddituum fructuum devoriarum proficuorum emolumentorum firmarum eleemozinarum lie daill sylver obituum et anniversariorum quorumcunque, que quovismodo pertinuerunt aut pertinere dinoscuntur ad quascunque

JAMES, by the grace of God King of Scots: To all good men of his whole land, clerics and laics, greeting. Know ye that we with the advice of the lords of our Privy Council have fully understood a certain charter and infeftment, made, given, and granted by our dearest mother, Queen of our realm for the time, after her perfect age, with the advice and consent of the lords of her Privy Council, to our lovites the Provost, Bailies, Councillors, and Community of our Burgh of Edinburgh and their successors, in regard to the gift, grant, and confirmation of all and sundry lands, tenements, houses, buildings, churches, chapels, yards, orchards, crofts, annual rents, fruits, duties, profits, emoluments, rents, alms, daill silver, obits, and anniversaries whatsoever, which any time belonged or are known to belong to any chaplainries, altarages, [or] prebends, founded or to be founded in any church, chapel, or college

capellanias alteragia prebendas in quacunque ecclesia capella aut collegio infra libertatem dicti Burgi nostri de Edinburgh, fundata seu fundanda per quemcunque patronum, in quarum possessione capellanii et prebendarii earundem perprius fuerunt, ubicunque prefate domus tenementa edificia pomaria horti annui redditus anniversaria fructus proventus et emolumenta jacent, aut prius leuata fuerunt respective, cum maneriebus locis hortis pomariis terris annuis redditibus emolumentis et devoriis quibuscunque que Fratribus Dominicalibus seu Predicatoribus et Minoribus seu Franciscanis dicti Burgi nostri de Edinburgh perprius pertinuerunt; unacum omnibus et singulis terris domibus tenementis et hortis jacentibus infra dictum nostrum Burgum et libertatem ejusdem, cum omnibus annuis redditibus de quacunque domo terris aut tenemento infra dictum nostrum Burgum leuandis, quibuscunque capellaniis alteragiis ecclesiis mortuariis aut anniversariis ubicunque sint infra regnum nostrum donatis dotatis et fundatis; Ac etiam cum omnibus et singulis annuis redditibus et aliis devoriis solitis, aut que per quamcunque ecclesiam extra dictum nostrum Burgum, a Preposito aut Ballivis ejusdem de communi redditu ejusdem pro suffragiis celebrandis demandari poterint, cum pertinentiis, ac de omnibus aliis privilegiis

within the liberty of our said Burgh of Edinburgh, by any patron, in the possession of which the chaplains and prebendaries of the same formerly were, wherever the said houses, tenements, buildings, orchards, yards, annual rents, anniversaries, fruits, profits, and emoluments are situated, or were formerly uplifted respectively, with the manor places, yards, orchards, annualrents, emoluments, and duties whatsoever, which formerly belonged to the Dominican or Preaching Friars and the Minorites or Franciscans of our said Burgh of Edinburgh; together with all and sundry lands, houses, tenements, and yards lying within our said Burgh and the liberty of the same, with all annualrents uplifted from any house, lands, or tenement, within our said Burgh, given, granted, and donated to chaplainries, altarages, churches, burials, or anniversaries, wherever they be within our kingdom; as also with all and singular annualrents and other duties customary, or that could be demanded by, any church outwith our said Burgh, from the provost or bailies of the same out of the common good of the same, for the celebration of suffrages, with the pertinents; and

libertatibus et facultatibus in carta et infeofamento donationis et dispositionis predictis desuper confectis ad longum specificatis et contentis tenendis de dicta charissima nostra matre et successoribus suis, de mandato nostro visam lectam inspectam et diligenter examinatam sanam integram non rasam non cancellatam nec in aliqua sui parte suspectam ad plenum intellexisse, sub hac forma :

MARIA Dei gratia Regina Scotorum: Omnibus probis hominibus totius terre sue clericis laicis, salutem [*etc. as in No. VIII.*].

QUAMQUIDEM cartam et infeofamentum in omnibus suis punctis et articulis conditionibus et modis ac circumstantiis suis quibuscunque in omnibus et per omnia forma pariter et effectu ut premissum est, approbamus ratificamus ac pro nobis et successoribus nostris pro perpetuo confirmamus. INSUPER nos cum avisamento predicto pro diversis rationabilibus causis bonis et considerationibus nos moventibus de novo tenore presentium damus concedimus et disponimus prefatis Preposito Ballivis Consulibus et Communitati dicti Burgi nostri de Edinburgh et eorum successoribus, omnes et singulas prenominatas terras tenementa domos edificia annuos redditus capellas loca hortos

of all other privileges, liberties, and faculties at length specified and contained in the said charter and infeftment of gift and disposition made thereupon, to be held of our said dearest mother and her successors,—by our command, seen, read, inspected, and diligently examined, perfect, whole, not erased, not cancelled, nor in any part suspect, in this form:

MARY, by the grace of God, Queen of Scots: To all good men of her whole land, clerics and laics, greeting [*etc. as above, No. VIII. p. 56*].

WHICH charter and infeftment in all its points and articles, conditions and modes and circumstances whatsoever, in all and by all, in the like form and effect, as premised, we approve, ratify, and for us and our successors confirm for ever. FURTHER we, with advice foresaid, for divers good and reasonable causes and considerations moving us, of new by the tenor of these presents, give, grant, and dispone to the foresaid Provost, Bailies, Councillors, and Community of our said Burgh of Edinburgh and their successors, all and sundry the before named lands, tenements, houses, buildings, annualrents, chapels, places, yards, orchards,

pomaria croftas census firmas proficua emolumenta et alia respective et particulariter superius specificata, per ipsos imperpetuum applicanda in sustentationem ministerii, pauperum auxilium, reparationem scolarum, propagationem literarum et scientiarum, pro eorum et successorum suorum arbitrio uti eis magis utile videbitur. Quibus etiam pro nobis et successoribus nostris plenariam ac liberam committimus potestatem quoscunque alios annuos redditus annua proficua quecunque tam extra quam intra dictum nostrum Burgum, que in posterum per quoscunque bono zelo ac liber[ali]tate sua motos ad alimentum ministrorum evangelii, auxilium pauperum, ac sustentationem gymnasiorum pro instaurandis scientiis et doctrina, donari et dotari contigerint acceptandi; Quas etiam terras annuos redditus et proficua suprascripta perprius donata et fundata ac in posterum donanda et fundanda ut premissum est, nos pro nobis et successoribus nostris nunc prout extunc et tunc prout exnunc confirmamus ratificamus et admortizamus ac easdem adeo libere mortificamus sicuti alique terre redditus tenementa et possessiones ecclesie ullo tempore precedenti mortificate fuerunt. Preterea nos pro nobis et successoribus nostris ratificamus approbamus et confirmamus

crofts, dues, rents, profits, emoluments, and others severally and particularly above specified, to be applied by them in all time coming to the sustentation of the ministry, the help of the poor, the repairing of schools, the propagation of letters and sciences, at the discretion of them and their successors as shall seem to them most advantageous. To whom also we, for ourselves and our successors, grant full and free power to accept whatever other annualrents and yearly profits, as well without as within our said Burgh, may in time coming happen to be given and doted by any persons, moved by good zeal and their own liberality, for the maintenance of the ministers of the gospel, the help of the poor, and sustentation of schools for the increase of science and learning; which lands, annualrents, and profits above written, formerly doted and founded and to be hereafter doted and founded as aforesaid, we, for us and our successors, now as then and then as now, confirm, ratify, and mortify, and the same we mortify as freely as any lands, rents, tenements, and possessions of the church were mortified in any time bygone. Moreover, we for us and our successors ratify, approve, and confirm the renunciation and demission made

renunciationem et dimissionem per familiarem servitorem nostrum Joannem Gib factam de omnibus jure et titulo que ipse virtute nostre donationis pretendere potuit ad preposituram Ecclesie beate Marie de Campis, (vulgo lie Kirk of Feild), cum fructibus terris possessionibus redditibus et devoriis ejusdem preteritis presentibus et futuris, in favorem dictorum Prepositi Ballivorum Consulum et Communitatis pro seipsis et eorum successoribus ac nomine et ex parte ministerii et pauperum. Ac quia intra privilegia et libertatem dicti nostri Burgi nunc diversa extant vasta et spatiosa loca que preposito prebendariis sacerdotibus et fratribus tempore preterito pertinuerunt maxime apta et commoda pro constructione domorum et edificiorum, ubi professores bonarum scientiarum et literarum ac studentes earundem remanere et suam diuturnam [diurnam] exercitationem habere poterint ultra et preter alia loca convenientia pro hospitalitate; Ideo nos, enixe cupientes ut in honorem Dei et commune bonum nostri regni literatura indies augeatur, volumus et concedimus quod licebit prefatis Preposito Consulibus et eorum successoribus edificare et reparare sufficientes domos et loca pro receptione habitatione et tractatione professorum scolarum grammati-

by our familiar servitor, John Gib, of all right and title to which he, by virtue of our gift, could pretend to the provostry of the Kirk of Saint Mary in the Fields, commonly called the Kirk of Field, with the fruits, lands, possessions, rents, and duties thereof, bygone, present, and to come, in favour of the said Provost, Bailies, Councillors, and Community, for themselves and their successors, and in name and on behalf of the ministers and the poor. And because there are now within the privileges and liberty of our said burgh divers waste and spacious places which formerly belonged to the provost, prebendaries, priests, and friars, very fit and commodious for the construction of houses and buildings where the professors and students of the liberal sciences and letters might stay and have their daily exercise, besides and beyond other places convenient for hospitality [or charity]. Therefore, we, earnestly desiring that for the honour of God and the common good of our kingdom literature should daily increase, will and grant that it shall be lawful to the said Provost, Councillors, and their successors, to build and repair sufficient houses and places for the reception, habitation, and entertainment of the professors of the schools of grammar, humanity, and the

calium, humanitatis, et linguarum, philosophie, theologie, medicine, et jurium, aut quarumcunque aliarum liberalium scientiarum, quo declaramus nullam fore rapturam predicte mortificationi; Ac etiam prefati Prepositus Ballivi et Consules ac eorum successores cum avisamento tamen eorum ministrorum pro perpetuo imposterum plenam habebunt libertatem personas ad dictas professiones edocendas maxime idoneas uti magis convenienter poterint eligendi cum potestate imponendi et removendi ipsos sicuti expediverit, ac inhibendo omnibus aliis ne dictas scientias intra dicti nostri Burgi libertatem profiteantur aut doceant nisi per prefatos Prepositum Ballivos et Consules corumque successores admissi fuerint; Proviso quod presentes nullatenus prejudicabunt nec actoribus nec reis nec aliis interesse habentibus in ejectione et causa prosequuta penes decimas garbales de Dumberny Potie et Moncreif ad capellanos Ecclesie Beati Egidii de Edinburgh pertinentes neque juri patronatus ejusdem, sed quod utrique parti et omnibus interesse habentibus usque ad finalem exitum et decisionem in hujusmodi ut congruit prosequi et defendere liceat presentibus aut quibuscunque in eisdem contentis non obstantibus. Proviso etiam quod ministri

languages, philosophy, theology, medicine, and law, or any other liberal sciences, whereby we declare there shall be no abstraction from the foresaid mortification. And also the said Provost, Bailies, and Councillors, and their successors, with advice, however, of their ministers, shall have full power in time coming to choose the most suitable persons as they can most conveniently for teaching the said professions, with power to place and remove them as shall be expedient; and discharging all others from professing or teaching the said sciences within the liberty of our said Burgh, unless they shall have been permitted to do so by the said Provost, Bailies, and Councillors and their successors. Providing that these presents should nowise prejudice either the pursuers or defenders or others having interest in the ejection and cause instituted anent the teind sheaves of Dumberny, Potie, and Moncreiff belonging to the chaplains of the church of Saint Giles of Edinburgh, nor the right of patronage to the same, but that it may be lawful to either party and all having interest to prosecute and defend the said pleas to the final issue and decision as in such case is meet, these presents, or anything contained in the same, notwithstanding. Providing also

deservientes apud dictas ecclesias pro presenti et in futurum sustentabuntur de promptioribus fructibus earundum secundum ordinem desuper sumptum seu sumendum. In cujus rei testimonium huic presenti carte nostre confirmationis magnum sigillum nostrum apponi precepimus. Testibus, predilectis nostris consanguineis et consiliariis Esmo Lennocie duce comite de Derulie domino Tarboltoun Dalkeith et Aubigny, &c., magno regni nostri camerario, Colino Argadie comite domino Campbell et Lorne, &c., cancellario ac justiciario nostro generali; reverendissimo ac venerabili in Christo patribus Patricio Sanctiandree archiepiscopo, Roberto commendatario monasterii nostri de Dunfermling, nostro secretario; dilectis nostris familiaribus et consiliariis Alexandro Hay, nostrorum rotulorum registri ac consilii clerico, Lodovico Bellenden de Auchnoule milite, nostre justiciarie clerico, Roberto Scott, nostre cancellarie directore, et magistro Thoma Buquhannane de Ybert, nostri secreti sigilli custode. Apud Castrum nostrum de Striviling, decimo quarto die mensis Aprilis, anno Domini millesimo quingentesimo octuagesimo secundo, regnique nostri anno decimo quinto.

that the ministers present and to come serving in the said churches, shall be sustained out of the readiest fruits of the same according to the orders made or to be made thereupon. In witness whereof we have caused our great seal to be affixed to this our present charter of confirmation. Witnesses, our well beloved cousins and councillors, Esme duke of Lennox earl of Darnley lord Tarbolton Dalkeith and Aubigny, etc. great chamberlain of our kingdom, Colin earl of Argyll lord Campbell and Lorne etc., our chancellor and justice general; the most reverend and venerable fathers in Christ, Patrick archbishop of St Andrews, Robert commendator of our monastery of Dunfermline, our secretary; our beloved servants and councillors, Alexander Hay, clerk of our rolls register and council, Louis Bellenden of Auchnoule knight, our justice clerk, Robert Scott, director of our chancery, and Master Thomas Buchanan of Ybert, keeper of our privy seal. At our castle of Stirling, the fourteenth day of the month of April, in the year of our Lord one thousand five hundred and eighty-two, and in the fifteenth year of our reign.

XII.

CONTRACT between the Provost, Bailies, Council, and Deacons of the Burgh of Edinburgh, and Mr Robert Pont, Provost of Trinity College, in regard to the renunciation of the Provostry. Edinburgh, 26th April 1585.

AT Edinburgh the xxvj day of Aprile, the yeir of God, j^{m} v^{c} four scoir fyve yeirs, it is appointit aggreit and finalie contractit betuix the honorable parteis following, to witt, the Provest, Bailleis, Counsale, and Deaconis of the Burgh of Edinburgh for thame and thair successouris on the ane parte, and Maister Robert Pont, provest of the Trinitie College beside Edinburgh, on that vther pairt, in maner eftir specifiit: That is to say, the said Maister Robert, movit be gude zeale, conscience, and eirnest affectioun to advance the Hospitallis and Colleges of the said Burgh, foundit or to be foundit be the saidis Provest, Balleis, and Counsale, and thair successouris within the samin for help and sustentatioun of the puir, seik, ageit, decrippit, faderles and orphenis, and for instructioun of the youth in letteres and virtew, quhairby cheritie may incresce to the glorie of God and his trew relligioun within this realme: Thairfore the said Maister Robert sall personalie be himselff or be his patent letteres of procuratioun seillit and subscriuit with his hand in dew forme, puirlie and simple dimitt, renunce, and resigne, lyke as the said Maister Robert be the tennour of this present contract, puirlie and simple dimittis, renunces, and resignis in the handis of our Soueraue Lord all and haill the said benefice of the Trinitie College beside Edinburgh, with all and sindrie kirkis, teyndschaves, vtheris teyndis, gleibis, manssis, biggingis, orcheardis, yairdis, annuelrentis, advocatioun, donatioun, and richt of patronage of prebendaries, chaiplainreis and donatioun of beidmenschippis, bedlyaris and vtheris offices pertening to the said Provestrie and Hospitall of the Trinitie College foundit beside the samin, togidder with the paroche kirk, personage, and vicarege of Sowtra and Lempetlaw and vtheris kirkis and teyndis annext to the said provestrie; and

with the place, orcheard, and yaird callit Dingwall Castell, pertening to the samin, and all and sindrie vtheris fructis, emolumentis, richtis, casualiteis, proffittis and dewiteis quhatsumeuir pertening and belanging to the said provestrie, quhaireuir the samin lyis within this realme, in favouris of the saidis Provest, Bailleis, Counsale, and Communitie of the said Burgh of Edinburgh, and thair successouris to remane with thame perpetualie in all tyme cuming, in puir and perpetuall almous, to be applyit and disponit be thame to the mantenance, help, and support of thair saidis hospitallis, college, and scuillis, the puir and scolleris of the samin, as thai sall think expedient, and as thai will ansuer to God at the lattir day; and all richt and titill of richt quhilk the said Maister Robert had, hes, or ony wyis may clame and haif to the said benefice and pertinents thairof forsaidis in tyme cuming, renunceand and dischargeand the samin for him and his successouris in fauouris and to the effect foirsaid for euir; And sall deliuer to the saidis Provest, Bailleis, and Counsale, the foundatioun, erectioun, charteris, sesingis, giftis and vtheris evidentis and writtis quhatsumeuir quhilk he has presentlie in his handis or salhappin heireftir to obtene, with the rentallis, decreittis, and letteres concerning the said provestrie, college, and hospitall, and sall mak or renew the said dimissioun and resignatioun at quhat tyme or howoft he salbe requirit thairto. And the said Maister Robert is content and consentis that the saidis Provest, Bailleis, and Counsale sall enter presentlie to the possessioun of the said college, hospitall, place, castell, houssis, biggings, yairdis, and pertinentis of the samin; with power to thame to mak and constitute bailleis, chalmerlains, factouris, maisteris of the hospitall, clerkis, seriandis, beddellis, and vtheris officieris neidfull, and to hald courte and courtis vpoun quhat place and als oft as thai sall think expedient; and to intromett and vptak the teyndis, fructis, males, fermes, annualrents, and vtheris emolumentis and dewiteis pertening to the said provestrie and hospital presentlie and in all tyme cuming, begynnand the first intromissioun at the said xxvj day of Apprill instant; and to do all vther thingis concerning the premissis, sielyke and als frelie as the said Maister Robert micht haue done befoir the making of this present contract. For the quhilkis caussis, and for divers vtheris gude deidis, gratitudis

and plesouris, done and schawin be the said Maister Robert for the weill of the said Burgh, and to the effect he sall nocht be preiugeit nor hurte in the yeirlie dewitie that he ressauit of the said benefice, or at the leist neir the valour thairof, the saidis Provest, Bailleis, and Counsale hes instantlie payit and deliuerit to the said Maister Robert the sowme of thre hundreth merkis vsuale money of this realme in contentatioun for all gressumes, entres syluer, and vtheris casualiteis quhilkis he mycht haif ressauit of the said benefice during his liftyme; of the quhilkis he haldis him weill content and payit and discharges thame thairof be thir presentis. And forder bindis and oblissis thame and thair successouris to content and thankfullie pay to the said Maister Robert yeirlie during all the dayis of his liftyme the soume of ane hundreth threscoir pundis money foirsaid at tua termes in the yeir, Witsounday and Martymes in winter be equale portiones, begynnand the first termes payment at the feist of Martymes nixttocum. And for the mair suir and thankfull payment of the said yeirlie dewitie to the said Maister Robert, the saidis Provest, Bailleis, and Counsale, for thame and thair successouris, bindis and oblissis thame within the space of ane moneth eftir the dimissioun of the said benefice, to infeft the said Maister Robert or ony vther he pleissis in his name, in ane annuelrent of ane hundreth threscoir pundis money foirsaid yeirlie, to be vpliftit during his liftyme at the termes foirsaidis, furth of thair commoun mylnis pertening to thair said toun of Edinburgh, sufficientlie be charter and sesing or at the leist be ane sufficient sesing to be given thairupoun; and to caus James Ros thair thesaurar present be actit in the commissaris buikis of Edinburgh for yeirlie payment of the said annuell salang as he bruikis the said office, and lykewyis thair thesauraris to cum within ane moneth eftir thair entre to the same actit as said is, to pay the said annuell during the tyme of thair offices bering respectiue and that during the said Maister Robertis liftyme. And the said Maister Robert binds and oblissis him to warrand and mak the yeirlie rent of the said benefice frelie to be worth yeirlie the said soume of ane hundreth threscoir pundis; and incais the samin benefice salbe of les availl heirefter be ony occasioun of his fact and deed or his predecessouris, in that cais the said Maister Robert bindis and oblissis him to defalk samelyke yeirlie of the soume

aboue specifiit quhilk the saidis Provost, Bailleis, and Counsale ar oblist to pay to him as said is, according as thai sal happin to want of the rentall and yeirlie dewitie of the said benefice through occassioun of the said Maister Robert or his predecessouris as is aboue specifiit; and sielyke to warrand the said dimissioun and resignatioun to be gude and sufficient in the selff to the effect foirsaid fra all richt and fact done be him in preiudice thairof. And for the mair securitie, bayth the saidis pairteis ar content that thir presentis be actit and registrat in the buikis of counsell, or commissaris, or townis buikis of Edinburgh, and thair autoritie to be interponit thairto with executorialis to be direct thairupoun in forme as efferis, and for the registering heirof makis and constitutis   thair lauchfull procuratouris in *communi forma promittendo de rato etc.* IN WITNES heirof bayth the saidis parties hes subscriuit thir presentis with thair handis, day, yeir, and place foirsaidis, befoir thir witnesses, Alexander Borthuik of Nether lany. Patrick Logy.

M. ROBERT PONT, prouest off the Trinitie College.
ALEXANDER BORTHUIK of Nethir lany.

XIII.

CHARTER by King James VI. to the Provost, Bailies, and Council of the Burgh of Edinburgh, of the Provostry of Trinity College. Dunfermline, 23d June 1585.

JACOBUS Dei gratia Rex Scotorum: Omnibus probis hominibus totius terre sue clericis et laicis salutem: Sciatis quod nos et domini nostri

JAMES, by the grace of God King of Scots: To all good men of his whole land clerics and laics, greeting: Know ye that we and the lords of our Privy Council

secreti consilii divinum animi zelum Prepositi Balliuorum et Consulum Burgi nostri de Edinburgh pro propagatione et decoratione suorum hospitalium collegiorum et ludorum literariorum infra dictum Burgum fundatorum considerantes, et hoc pro sustentatione pauperum et instructione juventutis in virtute humanisque literis, animoque voluentes nostro quam sit necessarium vti ipsis cum quodam patrimonio annuoque censu supportemus. Igitur nos cum auisamento predicto dedimus concessimus et in perpetuum mortificauimus disposuimus et confirmauimus tenoreque presentis carte nostre damus concedimus ac in perpetuum mortificamus disponimus et confirmamus dictis predilectis nostris Preposito Balliuis et Consulibus Burgi nostri de Edinburgh nunc presentibus et eorum successoribus totum et integrum beneficium prepositure Ecclesie Collegiate Trinitatis prope Edinburgh cum omnibus et singulis ecclesiis decimis garbalibus alijs decimis glebis mansis edificiis pomariis hortis annuis redditibus aduocationibus donationibus et jure patronatus prebendariorum et capellaniarum et donatione oratorum pauperum vocatorum vulgo beidmen et bedlyaris aliorumque officiorum dicte prepositure et Hospitalis Collegii Trinitatis, prope eundem fundatorum spectantibus

considering the holy zeal of the Provost Bailies and Councillors of our Burgh of Edinburgh for the extension and decoration of their hospitals colleges and grammar schools founded within the said Burgh, and that for the sustentation of the poor and instruction of youth in virtue and polite literature, and also pondering in our mind how necessary it is that we should support them with a certain patrimony and yearly income. Therefore we, with advice foresaid, have given, granted, and for ever mortified, disponed and confirmed, and by the tenor of our present charter, give, grant, and for ever mortify, dispone and confirm, to our said lovites the present Provost Bailies and Councillors of our Burgh of Edinburgh, and their successors, All and Whole the benefice of the provostry of the Collegiate Church of the Trinity near Edinburgh, with all and singular the churches, teind sheaves, and other teinds, glebes, manses, buildings, orchards, yards, annualrents, advocations, donations, and right of patronage of prebends and chaplainries, and presentation of poor orators, in Scots called beidmen and bedlyaris, and other officers of the said provostry and hospital of Trinity College founded near the same; together with the parish churches of

vnacum ecclesiis parochialibus de Sowtray et Lempetlaw aliisque ecclesiis ac decimis dicte prepositure annexatis cum loco pomario et horto vocato Dingwall Castell eidem spectantibus omnibusque aliis et singulis fructibus emolumentis juribus et casualitatibus proficuis devoriis tenentibus tenandriis et justis pertinentiis dicte prepositure spectantibus vbicumque infra regnum nostrum jacent; per dictos Prepositum Ballivos et Consules eorumque successores pro sustentatione seniorum decrepitorum orphanorum et pauperum infra dicta hospitalia ac pauperum scolasticorum infra dictum collegium et scolas omni tempore futuro intromittendis colligendis vtendis et disponendis, prout Deo Omnipotenti in extremo judicio respondere voluerint; Quodquidem beneficium prepositure Collegii Trinitatis cum omnibus et singulis pertinentiis eiusdem suprascriptis dilecto nostro oratori Magistro Roberto Pont vltimo preposito et possessori eiusdem perprius pertinuit, et per ipsum eiusque procuratores et patentes literas in manibus nostris ad effectum prescriptum per fustim et baculum apud Dunfermeling vicesimo tertio die mensis Junii instantis dimissum et resignatum fuerat; ac totum jus et jurisclameum proprietatem et possessionem que et quas in eodem

Soltray and Lempitlaw, and other churches and teinds annexed to the said provostry, with the place, orchard, and yard called Dingwall Castle belonging to the same, and all other and singular fruits, emoluments, rights, casualties, profits, duties, tenants, tenandries, and just pertinents belonging to the said provostry, wheresoever they lie within our kingdom: To be intromitted, ingathered, used, and disposed of by the said Provost Bailies and Councillors and their successors, for the sustentation of the aged, decrepit, orphans, and poor within the said hospitals, and of poor scholars within the said college and schools in all time coming, as they shall answer to God in the last judgment. Which benefice of the provostry of Trinity College, with all and sundry pertinents of the same above written, formerly belonged to our beloved orator Mr Robert Pont, the last provost and possessor thereof, and has been demitted and resigned by him and his procurators and letters patent in our hands to the effect foresaid, by staff and baton, at Dunfermline, on the twenty-third day of the current month of June, and he has for ever entirely upgiven all right and claim of right property and possession in the same which he has, had, or could

habuit habet seu quouismodo habere potuit omnino quiete clamauit in perpetuum. TENENDUM et HABENDUM totum et integrum dictum beneficium prepositure Ecclesie Collegiate Trinitatis cum omnibus et singulis ecclesiis decimis garbalibus alijsque decimis glebis mansis pomariis hortis annuis redditibus aduocationibus donationibus et jure patronatus prebendariorum capellaniarum et pauperum oratorum cum ecclesiis parochialibus rectoriis et vicariis de Soutra et Lempetlaw aliisque suprascriptis, dictis Preposito Balliuis et Consulibus eorumque successoribus ad effectum prescriptum in puram et perpetuam elemosinam de nobis et successoribus nostris in perpetuum per omnes rectas metas suas antiquas et diuisas prout jacent in longitudine et latitudine, in domibus edificiis boscis planis moris maresiis viis semitis aquis stagnis riuolis pratis pascuis et pasturis molendinis multuris et eorum sequelis aucupationibus venationibus piscationibus petariis turbariis carbonibus carbonariis cuniculis cunicnlariis columbis columbariis fabrilibus brasinis brueriis et genistis siluis nemoribus et virgultis lignis tignis lapicidiis lapide et calce, cum curiis querelis et earum exitibus herezeldis bludewitis adiudicamentis dictarumque curiarum eschcatis,

have in any manner of way. To HAVE and TO HOLD all and whole the said benefice of the provostry of the Collegiate Church of the Trinity, with all and sundry churches, teind sheaves and other teinds, glebes, manses, orchards, yards, annual rents, advocations, presentations and right of patronage of prebends chaplainries and poor beadsmen, with the parish churches parsonage and vicarage of Soltray and Lempetlaw and others above written, to the said Provost Bailies and Councillors and their successors, to the effect foresaid, in pure and perpetual alms, of us and our successors for ever, by all their just ancient bounds and marches as they lie in length and breadth, in houses, buildings, woods, plains, moors, marshes, ways, paths, waters, ponds, streams, meadows, pastures, mills, multures and their sequels, hawkings, huntings, fishings, peats, turfs, coals, coal pits, rabbits, rabbit-warrens, doves, dovecots, forges, malt-kilns, breweries, heaths, woods, groves, thickets, firewood, timber, stone quarries, stone and lime, with courts suits and their issues, herezelds, bludewites, adjudgments and escheats of the said courts, with common pasturage, free ish and entry, and with all other and singular liberties, commodities, profits and easements and their just pertinents whatsoever, as

cum communi pastura libero introitu et exitu, ac cum omnibus aliis et singulis libertatibus commoditatibus proficuis et asiamentis ac justis suis pertinentiis quibuscunque, tam non nominatis quam nominatis tam subtus terra quam supra terram, procul et prope ad predictum beneficium preposituro antedicte cum omnibus et singulis eiusdem pertinentiis spectantibus seu juste spectare valentibus quomodolibet, in futurum adeo libere quiete plenarie integre honorifice bene et in pace, in omnibus et per omnia, sicuti dictus magister Robertus aut alii sui predecessores dicto beneficio gauisi sunt, absque reuocatione contradictione aut obstaculo quocunque. Faciendo inde annuatim dicti Prepositus Balliui Consules et Communitas dicti Burgi pauperesque dicti hospitalis scolasticique dicti collegii et scolarum eorumque successores deuotas et humiles quotidianas preces Deo Omnipotenti pro preseruatione nostri successorumque nostrorum ac sustentando ministros curam dictarum ecclesiarum seruientes dicte prepositure spectantium eorumque successores, vel soluendo tertiam partem fructuum dicte prepositure pro eorum sustentatione ad eorum optionem et electionem tantum: Insuper cum potestate dilectis nostris Magistro Joanni Craig verbi Dei ministris eorumque alicui conjunctim et diuisim quatenus institutionem et

well named as not named, as well above the earth as below the earth, far and near, belonging or that could justly belong in any manner of way, to the foresaid benefice of the foresaid provostry, with all and singular the pertinents of the same for ever, as freely quietly fully wholly honourably well and in peace, in all and by all, as the said Master Robert or others his predecessors enjoyed the said benefice, without revocation contradiction or obstacle whatsoever. Rendering therefore yearly, the said Provost Bailies Councillors and Community of the said Burgh, and the poor of the said hospital, and scholars of the said college and schools, and their successors, devout and humble daily prayers to God Almighty for the preservation of us and our successors, and sustaining the ministers serving the cure of the said churches belonging to the said provostry, and their successors, or paying the third part of the fruits of the said provostry for their sustentation, at their option and choice allenarly. Moreover, with power to our lovites Master John Craig ministers of God's word, and to any one of them conjunctly and severally, that ye give and deliver, or that any one of you give and deliver,

possessionem dicte beneficii dictis Preposito Balliuis et Consulibus dicti nostri Burgi de Edinburgh vel eorum certo actornato latori presentium secundum tenorem presentis carte nostre tradatis et deliberetis seu aliquis vestrum tradat et deliberet sine dilatione et hoc nullo modo omittatis. In cuius rei testimonium huic presenti carte nostre magnum sigillum nostrum apponi precepimus. Testibus, predilecto nostro consanguineo et consiliario Jacobo Arranie comite domino Evane et Hamiltoun etc. cancellario nostro; reuerendissimo ac venerabili in Christo patribus, Patricio Sanctiandree archiepiscopo, Waltero commendatario prioratus de Blantyre nostri secreti sigilli custode; dilectis nostris familiaribus et consiliariis domino Joanne Maitland de Thirlstane milite nostro secretario, Alexandro Hay de Eister Kennat nostrorum rotulorum registri ac consilii clerico, Lodonico Bellenden de Auchnoule milite nostre justiciarie clerico, et Roberto Scott nostre cancellarie directore; Apud Dunfermeling vicesimo tertio die mensis Junij anno Domini millesimo quingentesimo octuagesimo quinto et regni nostri decimo octauo.

institution and possession of the said benefice to the said Provost Bailies and Councillors of our said Burgh of Edinburgh, or to their certain attorney, bearer of these presents, according to the tenor of our present charter, without delay, and this in noways ye leave undone. In witness whereof we have commanded our great seal to be affixed to this our present charter. Witnesses our well beloved cousin and councillor James earl of Arran lord Evane and Hamilton etc., our chancellor; the most reverend and venerable fathers in Christ, Patrick archbishop of St Andrews, Walter commendator of the priory of Blantyre, keeper of our privy seal; our beloved servants and councillors, Sir John Maitland of Thirlstane knight, our secretary, Alexander Hay of Easter Kennet, clerk of our rolls register and council, Louis Bellenden of Auchnoule knight, our justice-clerk, and Robert Scott director of our chancery; At Dunfermline the twenty-third day of the month of June in the year of our Lord one thousand five hundred and eighty-five, and the eighteenth of our reign.

XIV.

CHARTER by King James VI., confirming his previous Charter of 23d June 1585, and of new granting Trinity College and the whole endowments and property thereof, to the Provost, Bailies, Council, and Community of the Burgh of Edinburgh. Holyrood, 26th May 1587.

JACOBUS Dei gratia Rex Scotorum: Omnibus probis hominibus totius terre sue clericis et laicis, salutem: SCIATIS nos cum auisamento et consensu dominorum nostri secreti consilii quandam donationem dispositionem et mortificationem per nos factam per nostram cartam nostro sub magno sigillo de data apud Dunfermeling vicesimo tertio die mensis Junij Anno Domini millesimo quingentesimo octuagesimo quinto dilectis nostris Preposito Balliuis Consulibus et Communitati Burgi nostri de Edinburgh et eorum successoribus de toto et integro beneficio prepositure Ecclesie Collegiate Trinitatis prope Edinburgh cum omnibus et singulis ecclesiis decimis garbalibus aliis decimis glebis mansis edificiis pomariis hortis annuis redditibus advocationibus donationibus et jure patronatus prebendariorum et capellaniarum dicti Collegij, ac cum donatione oratorum pauperum vulgo vocatorum beidmen et bedlyaris aliorumque

JAMES, by the grace of God King of Scots: To all good men of his whole land, clerics and laics, greeting: Know ye that we, with the advice and consent of the lords of our Privy Council, have fully understood a certain gift, disposition, and mortification made by us by our charter, under our great seal, dated at Dunfermline on the twenty-third day of the month of June, in the year of our Lord one thousand five hundred and eighty-five, to our lovites the Provost, Bailies, Councillors, and Community of our Burgh of Edinburgh, and their successors, of All and Whole the benefice of the provostry of the Collegiate Church of the Trinity, near Edinburgh, with all and sundry churches, teind sheaves, and other teinds, glebes, manses, buildings, orchards, yards, annual rents, advocations, donations, and right of patronage of the prebends, and chaplainries of the said College, and with the presentation of poor orators, in Scots called beidmen and bedlyaris, and other officers of the said provostry

officiorum dicte prepositure et hospitalis Collegii Trinitatis prope eundem fundatorum spectantibus, vnacum ecclesiis parochialibus de Sowtray et Lempitlaw aliisque ecclesiis et decimis dicte prepositure annexatis, cum loco pomario et horto vocato Dingwall Castell eidem spectantibus, omnibusque aliis et singulis fructibus emolumentis juribus casualitatibus proficuis devoriis tenentibus tenandriis et justis pertinentiis dicto prepositure spectantibus, vbicunque infra regum nostrum jacent, de mandato nostro visam lectam inspectam et diligenter examinatam, sanam integram non rasam non cancellatam nec in aliqua sui parte suspectam ad plenum intellexisse sub hac forma:—

JACOBUS Dei gratia Rex Scotorum: Omnibus probis hominibus totius terre sue clericis et laicis, salutem [*etc. as above No. XIII.*]:

Quamquidem donationem dispositionem et mortificationem in omnibus suis punctis et articulis conditionibus et modis ac circumstantiis suis quibuscunque, in omnibus et per omnia forma pariter et effectu vt premissum est, approbamus ratificamus ac pro nobis et successoribus nostris pro perpetuo confirmamus: Insuper nos pro bono fideli et gratuito seruitio nobis et nostris predecessoribus bone memorie per dictos Prepositum Balliuos Consules et Communitatem dicti nostri Burgi eorumque

and Hospital of Trinity College, founded near the same, together with the parish churches of Soltray and Lempitlaw, and other churches and teinds annexed to the said provostry, with the place, orchard, and yard called Dingwall Castle, belonging to the same, and all other and singular fruits, emoluments, rights, casualties, profits, duties, tenants, tenandries, and just pertinents belonging to the said provostry wheresoever they lie within our realm,—by our command read, inspected, and diligently examined, whole, entire, not erased, not cancelled nor suspected in any part, in this form:—

JAMES, by the grace of God King of Scots: To all good men of his whole land, clerics and laics, greeting [*etc. as above, No. XIII. p. 83*].

Which gift, disposition, and mortification in all its points, articles, conditions, and modes and circumstances whatsoever, and in all and by all, in the like form and effect as aforesaid, we approve, ratify, and for us and our successors for ever confirm. Moreover we, for the good, faithful, and free service rendered and performed to us and our predecessors of happy memory by the said Provost, Bailies, Councillors, and Community of our said Burgh, and their pre-

predecessores omnibus temporibus retroactis prestito et impenso; Ac considerantes bonum et diuinum animi zelum quem erga sustentationem ministrorum evangelii inde residentium habent et gerunt et [qui] postea Deo volente infra dictum nostrum Burgum remanebunt, ac etiam vt studia humanarum literarum infra idem florere et increscere possunt et quod per hospitalium sustentationem indigentes morboque laborantes confortari possunt; super quibus diuinis respectibus et causis et alijs ad publicam honestatem decorationemque dicti nostri Burgi tendentibus pro reipublicce eiusdem propagatione burgum principale regni nostri existentis, vbi nos nostrique tres status regni sepissime residentiam habemus, dicti Prepositus Balliui Consules et Communitas magnas pecuniarum summas hactenus contulerunt, et absque nostro rationabili juvamine et supportatione suas diuinas animi intentiones perficere et complere non sunt habiles nec personas officia ministrorum collegii et ludorum literariorum gerentes cum pauperibus imbecillibus impotentibusque sustentare possunt; Ideo nos nobiscum plene resoluti maturaque deliberatione et auisamento prehabita omnes fructus proficua et emolumenta dicti collegii Collegii Trinitatis nuncupati permutare tam ad prepositum

decessors, in all time bygone; and considering the good and godly zeal which they have and bear towards the sustaining of the ministers of the gospel now residing, and who by the will of God shall afterwards reside, in our said Burgh, as also that the studies of polite letters may flourish and increase within the same, and that by the upholding of hospitals the poor and those labouring under disease may be comforted; Upon which godly accounts and causes, and others tending to the public credit and decoration of our said Burgh, and for the advancement of the commonweal of the same, being the principal burgh of our kingdom, where we and the three estates of our realm very often reside, the said Provost, Bailies, Councillors, and Community have heretofore contributed great sums of money, and without our reasonable help and support they are not able to perfect and complete their pious intentions, nor can they sustain the persons filling the offices of the ministry, college, and grammar schools, with the poor, imbecile, and impotent. Therefore we, after mature deliberation and advice, being fully resolved with ourselves to alter the destination of the whole fruits, profits, and emoluments of the said College called Trinity College, as well those belonging and pertaining to the provost as to the

quam ad prebendarios capellanos et alia eiusdem membra spectantia et pertinentia, seruitia pro quibus hujusmodi fundate fuere prius nunc minime necessaria existentia et easdem in usum ministrorum professorum literarum et pauperum sustentationem transferre cum auisamento et consensu antedicto de nouo dedimus concessimus et mortificauimus tenoreque presentis carte nostre damus concedimus et mortificamus prefatis Preposito Balliuis Consulibus et Communitati Burgi nostri de Edinburgh antedicti eorumque successoribus in perpetuum ad vsus subscriptos tantummodo, totam et integram predictam prepositurам Trinitatis Collegii cum omnibus et singulis ecclesiis decimis garbalibus ac aliis decimis glebis mansis edificiis pomariis hortis annuis redditibus aduocationibus donationibus ac jure patronatus prebendariorum et capellaniarum dicti Collegii cum donatione oratorum pauperum vulgo beidmen et bedlyaris nuncupatorum, aliorumque officiorum dicte prepositure et Hospitalis dicti Collegii prope eandem fundatorum spectantibus, unacum ecclesiis parochialibus de Sowtraw et Lempitlaw aliisque ecclesiis et decimis dicte prepositure annexatis cum loco pomario et horto vocato Dingwall eisdem spectantibus omnibusque aliis et singulis fructibus emolumentis juribus casualitatibus proficuis deuoriis tenentibus ten-

prebendaries, chaplains and other members thereof, the services for which these were formerly founded being now nowise necessary, and to transfer the same to the use of the ministers, the teaching of literature, and the sustaining of the poor, with advice and consent foresaid, of new have given, granted, and mortified, and by the tenor of our present charter, give, grant, and mortify to the said Provost, Bailies, Councillors, and Community of our said Burgh of Edinburgh and their successors for ever, for the uses underwritten only, All and Whole the foresaid provostry of Trinity College, with all and sundry churches, teind sheaves, and other teinds, glebes, manses, buildings, orchards, yards, annual rents, advocations, donations, and right of patronage of prebends, and chaplainries of the said College, with the presentation of poor orators in Scots called beidmen and bedlyaris, and other officers of the said provostry and Hospital of the said College, founded near the same, together with the parish churches of Soltray and Lempitlaw, and other churches and teinds annexed to the said provostry, with the place, orchard, and yard called Dingwall, belonging to the same, and all other and singular fruits, emoluments, rights,

andriis et justis pertinentiis dicte prepositure spectantibus vbicumque infra regnum nostrum ad burgum seu terras jacent. Ac cum omnibus et singulis ecclesiis decimis fructibus deuoriis emolumentis annuis redditibus et proficuis quibuscunque ad omnia et singula prebendaria et capellania dicti Collegii aut ad singulos predictos prebendarios in communitate, seu alicui vni eorum in proprietate spectantibus, aut per ipsos ipsorumve aliquem possessis antea, cum omnibus redditibus proficuis emolumentis terris et tenementis ad prefatum Collegium prepositum prebendarios et membra eiusdem vel ad sustentationem ecclesie domorum edificiorumque dicti Collegii fundatis et mortificatis, cum potestate dictis Preposito Balliuis et Communitati ac Consulibus eorumque successoribus leuandi recipiendi et intromittendi per seipsos eorum factores et procuratores eorum nomine omnes et singulos fructus proficuos et emolumenta predicta ac hujusmodi ad ministrorum sustentationem Collegii ludorum literarum et pauperum secundum eorum bonam discretionem super quam eorum conscientiam oneramus applicandi; Necnon volumus et concedimus tenoreque presentis carte nostre decernimus et ordinamus quod prefati Prepositus Balliui Consules et Communitas eorumque suc-

casualties, profits, duties, tenents, tenandries, and just pertinents belonging to the said provostry, wheresoever they lie within our realm, to burgh or to land, and with all and sundry churches, teinds, fruits, duties, emoluments, annualrents, and profits whatsoever belonging to all and sundry the prebends and chaplainries of the said College, or to each of the said prebendaries in common, or to any one of them in particular, or formerly possessed by them or any one of them, with all the rents, profits, emoluments, lands, and tenements founded and mortified to the foresaid College, provost, prebendaries, and members of the same, or to the upholding of the church, houses, and buildings of the said College. With power to the said Provost, Bailies, and Community, and Councillors, and their successors, to uplift, receive, and intromit with, by themselves or their factors and procurators in their name, all and sundry the foresaid fruits, profits, and emoluments, and to apply the same to the sustaining of the Ministers, College, Grammar Schools, and Poor, at their own good discretion, whereanent we burden their consciences. As also we will and grant, and by the tenor of our present charter decern and ordain, that the said Provost, Bailies, Councillors, and Community, and their successors, shall nowise be bound or obliged, any clauses contained in

cessores ad aliquem prebendarium seu capellanum ad prebendaria seu capellania dicti collegij nunc vacantia seu que postea quando vacare contigerit aut contigerint aliquodve particulare titulum ipsis hujusmodi concedendum, minime astricti aut obligati erunt, quibusvis clausulis in dicta fundatione contentis non obstantibus, quas tenore presentis carte nostre desoluimus et abrogamus vt hec presens nostra mortificatio validum effectum capiat, et quod predicta proficua simul inuicem collecta et congregata in vno rentale ad vsus prescriptas disponantur; preterea nos cum auisamento predicto volumus et concedimus quod prefati Prepositi Balliui Consules et Communitas dicti nostri Burgi eorumque successores plenum jus proprietatis habent et omni tempore futuro habebunt in et ad omnes et singulas terras tenementaque ad predictum collegium prepositum prebendarios hospitalarios et membra eiusdem annexata seu spectantia, et similiter ad superioritatem omnium terrarum de ipsis eorumve aliquo tentarum, feodatarios et alios tenentes earundem intrandi, firmas et deuorias per ipsos debitas leuandi, pro reductione recognitione et nonintroitu citandi, simili modo sicuti aliqui alij superiores per leges nostri regni fecerunt seu facere possunt, et sicuti dicti prepositus prebendarii et hospitalares ratione eorum fundationis vel alias

the said foundation notwithstanding, to present any prebendary or chaplain to the prebends or chaplainries now vacant, or that may hereafter become vacant, nor to grant to them any special title to the same; which clauses we by the tenor of these presents annul and abrogate that this our present mortification may receive valid effect, and that the foresaid profits may be all collected and gathered together in one rental and disponed to the foresaid uses. Besides we, with advice foresaid, will and grant that the foresaid Provost, Bailies, Councillors, and Community of our said Burgh, and their successors, have and shall have the full right of property, for all time coming, in and to all and sundry lands and tenements annexed or belonging to the foresaid College, provost, prebendaries, hospitallers, and members of the same, and likewise to the superiority of all the lands holden of them or any of them, of entering feuars and others tenants of the same, of uplifting the rents and duties due by them, and of summoning for reduction, recognition, and non-entry in the same manner as any other superiors by the laws of our kingdom have done or may do, and as the said provost, prebendaries, and hospitallers did or might have done by reason of

ullo tempore preterito fecerunt seu facere potuerunt, Et quia domus dicti hospitalis Trinitatis Collegii nuncupati nunc ruinosa extat et absque magnis expensis minime reparari potest, et eiusdem reparatio nunc minime est necesse quia dicti Prepositus Balliui Consules et Comunitas hospitale magis idoneum in vna parte dicte Ecclesie Collegiate situatum magis aptum et conueniens quam dictum vetus hospitale fuit construxerunt et reparauerunt, ac illud cum sufficienti furnitura et necessariis pro asiamento pauperum et morbis laborantium in eadem recipiendorum, nos pro nobis et successoribus nostris volumus et concedimus quod licebit prefatis Preposito Balliuis Consulibus et Communitati eorumque successoribus tot pauperes infra Hospitale per ipsos nuper reparatim sustentare, ut per eos super redditibus dicti hospitalis Trinitatis Collegii conuenienter sustentari possunt, pro quibus tenore presentis carte nostre obligati et astricti erunt; necnon dictum netus ruinosum Hospitale quocunque profitabili vsui ipsis magis videbitur expediens applicare; Tenendam et Habendam totam et integram predictam preposituram Trinitatis Collegii cum omnibus et singulis ecclesiis decimis garbalibus et aliis decimis glebis mansis edificiis pomariis hortis annuis

their foundation or otherwise any time bygone. And because the house of the said Hospital called Trinity College is now ruinous and cannot be repaired in anywise without great expenses, and the repairing thereof is now in no wise necessary, because the said Provost, Bailies, Councillors, and Community have built and repaired a more suitable Hospital, situated in a part of the said Collegiate Church, more fit and convenient than the said old Hospital was, and have provided the same with sufficient furniture and necessaries for the relief of the poor and those labouring under disease to be received into the same, we, for us and our successors, will and grant that it shall be lawful to the said Provost, Bailies, Councillors, and Community, and their successors to sustain as many poor within the Hospital lately repaired by them as may be conveniently maintained upon the rents of the said Hospital of Trinity College, for which, by the tenor of this our present charter, they shall be obliged and astricted; as also to apply the said old ruinous Hospital to whatever profitable use shall seem to them most expedient. To hold and have all and whole the foresaid provostry of Trinity College, with all and sundry churches, teind sheaves, and other teinds, glebes, manses,

redditibus aduocationibus donationibus et jure patronatus prebendariorum et capellaniarum dicti Collegii cum donatione oratorum pauperum vulgo beidmen et bedlyaris nuncupatorum aliorumque officiorum dicti prepositure et Hospitali dicti Collegii prope eandem fundatorum spectantibus vnacum dictis ecclesiis parochialibus de Sowtray et Lempitlaw aliisque ecclesiis et decimis dicte prepositure annexatis cum prenominato loco pomario et horto vocato Dingwall Castell eidem spectantibus, omnibusque aliis et singulis fructibus emolumentis juribus proficuis deuoriis tenentibus tenandriis ac justis pertinentiis dicte prepositure spectantibus vbicunque infra dictum regnum nostrum ad burgum seu terras jacent, ac cum omnibus et singulis dictis deuoriis decimis fructibus ecclesiis emolumentis annuis redditibus et proficuis quibuscunque ad omnia et singula dicta prebendaria et capellania dicti Collegii aut ad singulos predictos prebendarios in communitate seu alicui eorum vni in proprietate spectantibus, aut per eos ipsorumque aliquem antea possessis, cum omnibus redditibus proficuis emolumentis terris et tenementis ad predictum Collegium prepositum prebendarios et membra eiusdem vel ad sustentationem ecclesie domorum edificiorumque

buildings, orchards, yards, annual rents, advocations, donations, and right of patronage of prebendaries, and chaplains of the said College, with the presentation of poor orators in Scots called beidmen and bedlyaris, and other officers of the said provostry and Hospital of the said College, founded near the same, together with the said parish churches of Soltray and Lempitlaw, and other churches and teinds annexed to the said provostry, with the forenamed place, orchard, and yard called Dingwall Castle, belonging to the same, and all other and sundry fruits, emoluments, rights, profits, duties, tenants, tenantries, and just pertinents belonging to the said provostry, wheresoever situated within our kingdom, to burgh or to land; and with all and sundry the said duties, teinds, fruits, churches, emoluments, annual rents, and profits whatsoever belonging to all and sundry the said prebends and chaplainries of the said College, or to each of the foresaid prebendaries in common, or to any one of them in particular, or previously possessed by them or any of them; with all the rents, profits, emoluments, lands, and tenements founded and mortified to the said College, provost, prebendaries, and members of the same,

dicti Collegii fundatis et mortificatis, prefatis Preposito Balliuis Consulibus et Comuunitati dicti nostri Burgi de Edinburgh eorumque successoribus, ad vsus et effectum suprascriptos solummodo, de nobis et nostris successoribus in pura et perpetua elemosina in perpetuum per omnes rectas metas suas antiquas et diuisas prout jacent in longitudine et latitudine in domibus edificiis boscis planis moris marcsiis viis semitis aquis stagnis riuolis pratis pascuis et pasturis molendinis multuris et eorum sequelis aucupationibus venationibus piscationibus petariis turbariis carbonibus carbonariis cuniculis cunicularìis columbis columbariis fabrilibus brasinis brueriis et genestis siluis nemoribus et virgultis lignis tignis lapicidiis lapide et calce cum curiis et earum exitibus herezeldis bludewitis et mulierum marchetis cum communi pastura liberoque introitu et exitu ac cum omnibus aliis et singulis libertatibus commoditatibus proficuis et asiamentis ac justis suis pertinentiis quibuscunque, tam non nominatis quam nominatis tam subtus terra quam supra terram procul et prope ad predictam prepositurain cum vniuersis et singulis ecclesiis decimis et deuoriis antedictis cum suis pertinentiis particulariter prescriptis spectantibus seu juste spectare

or to the maintenance of the church houses and buildings of the said College to the foresaid Provost, Bailies, Councillors, and Community of our said Burgh of Edinburgh, and their successors, to the uses and effect above written only, of us and our successors in pure and perpetual alms for ever, by all their just and ancient marches and divisions as they lie in length and breadth, in houses, buildings, woods, plains, moors, marshes, ways, paths, waters, ponds, streams, meadows, pastures and feeding grounds, mills, multures, and their sequels, hawkings, huntings, fishings, peat and turf, coals, coal pits, rabbits, rabbit-warrens, doves, dovecots, forges, malt kilns, breweries, heaths, woods, groves and thickets, firewood, timber, stone quarries, stone and lime, with courts and their issues, herezelds, bludewites, marchets of women, with common pasturage, and free ish and entry, and with all other and singular liberties, commodities, profits, and easements, and their just pertinents whatsoever, as well not named as named, as well below the earth as above, far and near, belonging or that could justly belong in any manner of way to the foresaid provostry, with all and sundry churches, teinds, and duties aforesaid, with their pertinents particularly

valentibus quomodolibet, in futurum libere quiete plenarie integre honorifice bene et in pace absque ulla reuocatione contradictione impedimento aut obstaculo quocunque; Faciendo inde annuatim dicti Prepositus Balliui et Communitas dicti nostri Burgi pauperesque dicti hospitalis scolasticique dicti collegii et scolarum eorumque successores deuotas et humiles quotidianas preces Deo Omnipotenti pro preseruatione nostri successorumque nostrorum ac sustentatione ministrorum curam dictarum ecclesiarum seruientium dicto preposituro spectantium eorumque successores, vel soluendo tertiam partem fructuum dicti prepositure pro eorum sustentatione ad eorum optionem et electionem; Ac preterea quod prefati Prepositus Balliui Consules et Communitas eorumque successores omnes fructus annuos redditus et proficua prescripta ad vsus predictos impensare et conferre astricti et obligati erunt; et quod nobis et successoribus nostris pro hujusmodi computabiles existent quandocunque requisiti fuerint; reseruando nihilominus totis prebendariis dicti Collegii adhuc viuentibus tantas deuorias annuatim ut vnusquisque eorum recipere consueuit, de quibus deuoriis prefati Prepositus et Balliui illis responderi facientur durante eorum vita tantum. In cujus rei testimonium

before written, freely, quietly, fully, wholly, honourably, well and in peace in all time coming, without any revocation, contradiction, impediment, or obstacle whatsoever. Giving therefor annually the said Provost, Bailies, and Community of our said Burgh, and the poor of the said Hospital, and scholars of the said College and Schools, and their successors, devout and humble daily prayers to Almighty God for the preservation of us and our successors, and the sustaining of the ministers serving the cure of the said churches belonging to the said provostry, and their successors, or paying the third part of the fruits of the said provostry for their sustenance, at their option and choice. And, moreover, that the foresaid Provost, Bailies, Councillors, and Community, and their successors shall be bound and obliged to lay out and expend all the foresaid fruits, annual rents, and profits to the foresaid uses; and that they shall be accountable to us and our successors for the same whenever they shall be required; Reserving nevertheless to all the prebendaries of the said College at present living so much of the said duties yearly as each of them was in use to receive, for which duties the said Provost and Bailies shall be made answerable to them during their lifetime only:

huic presenti carte nostre confirmationis magnum sigillum nostrum apponi precepimus: Testibus, predilectis nostris consanguineis et consiliarijs Joanne domino Hamiltoun etc. comendatario monasterii nostri de Abirbrothok, Archibaldo Angusie comite, domino Dowglas Dalkeyth et Abirnethie, reuerendissimo ac venerabili in Christo patribus Patricio Sancti Andree archiepiscopo, Waltero priore de Blantyre nostri secreti sigilli custode; dilectis nostris familiaribus et consiliariis domino Joanne Maitland de Thirlstane milite nostro secretario, Alexandro Hay de Eister Kennat nostrorum rotulorum registri ac consilii clerico, Lodouico Bellenden de Auchnoule milite nostre justiciarie clerico, et magistro Roberto Scott nostre cancellarie directore, Apud Halierudehous vicesimo sexto die mensis Maij anno Domini millesimo quingentesimo octuagesimo septimo et regni nostri vicesimo.

In witness whereof we have commanded our great seal to be affixed to this our present charter of confirmation: Witnesses, our well beloved cousins and councillors, John lord Hamilton, etc. commendator of our monastery of Arbroath, Archibald earl of Angus lord Douglas Dalkeith and Abernethy; the most reverend and venerable fathers in Christ, Patrick archbishop of St Andrews, Walter prior of Blantyre, keeper of our privy seal; our beloved servants and councillors, Sir John Maitland of Thirlstane knight, our secretary, Alexander Hay of Easter Kennet, clerk of our rolls register and council, Louis Bellenden of Auchnoule knight, our justice clerk, and Master Robert Scott, director of our chancery: At Holyroodhouse, the twenty-sixth day of the month of May, in the year of our Lord one thousand five hundred and eighty-seven, and the twentieth of our reign.

XV.

CHARTER by King James VI., confirming his Charter of 23d June 1583, and of new granting to the Provost, Bailies, Council, and Community of the Burgh of Edinburgh, the whole revenues of Trinity College. Holyrood, 29th July 1587.

JACOBUS Dei gratia Rex Scotorum: Omnibus probis hominibus totius terre sue clericis et laicis, salutem: SCIATIS quia nos post nostram perfectam et legitimam etatem viginti vnius annorum completam in parliamento nostro declaratam, et generalem reuocationem nostram in hujusmodi factam, nunc moti ardenti zelo et diuina intensione Prepositi Balliuorum Consulum et Communitatis Burgi nostri de Edinburgh in suorum ministrorum euangelii prouisione infra dictam nostram Burgum seruientium qui nulla certa aut constituta stipendia ex nostris tertijs beneficiorum habent; et quod ipsi magnas pecuniarum summas pro edificatione vnius hospitalis vbi Collegium Reginale vulgo the Quenis College perprius stetit pro pauperum et miserabilium personarum sustentatione contulerunt, et quod preterea Collegium infra dictum nostrum Burgum nuper erexerunt in quo bone litere scientieque professe sunt pro regni

JAMES, by the grace of God King of Scots: To all good men of his whole land, clerics and laics, greeting: Know ye that we, after our perfect and lawful age of twenty-one years complete declared in our parliament, and our general revocation made in the same, now moved by the ardent zeal and godly purpose of the Provost, Bailies, Councillors and Community of our Burgh of Edinburgh in providing for their Ministers of the gospel serving within our said Burgh who have no certain or ascertained stipends out of our thirds of benefices; and that they have contributed large sums of money for the building of an Hospital where the Queen's College formerly stood, for the sustentation of poor and miserable persons; and that besides they have lately erected a College within our said Burgh, in which polite letters and sciences are taught, for the benefit of the kingdom, and for the public credit and decoration of our

commoditate et ad publicam honestatem et decorationem dicti nostri Burgi pro re[i] publice eiusdem propagatione burgum principale huius nostri regni existentis vbi nos nostrique tres status regni sepissime residentiam habemus; Quiquidem ministri hospitale et collegium antedicta absque nostro rationabili juvamine et supportatione minime bene sustentari possunt non obstantibus magnis expensis tam de communi bono dicti nostri Burgi quam de particulari contributione ex his qui ad juvamen et supplementum dicti diuini operis se prebuerunt per predictos Prepositum Balliuos Consules et Communitatem hactenus desuper confecti. Et considerantes nos nostramque quondam charissimam matrem diuersas terras redditus decimas et annuos redditus ad sustentationem dicti ministerii hospitalis et collegii dotasse concessisse et mortificasse quas volumus cum prefatis Preposito Balliuis Consulibus Communitati et eorum successoribus in perpetuum remanere, Et intelligentes easdem minime sub annexatione terrarum ecclesiasticarum nostre corone comprehendi, et quod e nostra generali reuocatione nuper facta excepte sunt; IDEO ratificauimus approbauimus ac pro nobis et successoribus nostris pro perpetuo confirmauimus tenoreque presentis carte

said burgh, for the advancement of the commonweal of the same, being the principal burgh of this our kingdom, where we and the three estates of the realm very often reside; which Ministers, Hospital, and College foresaid could not be well sustained without our reasonable help and support, notwithstanding the great expenses, as well from the common good of our said Burgh, as from the particular contributions of those who have devoted themselves to the support and aid of the said pious work thus far executed thereanent by the said Provost, Bailies, Councillors, and Community. And considering that we and our late dearest mother have given granted and mortified divers lands, rents, teinds, and annual rents for the sustentation of the said Ministry, Hospital, and College, which we wish to remain with the said Provost, Bailies, Councillors, Community, and their successors for ever. And understanding the same not to be comprehended in the annexation of the church lands to our Crown, and that they are excepted from our general revocation lately made, Therefore, we have ratified, approved, and for us and our successors, confirmed for ever, and by the tenor of our present charter, ratify, approve, and

nostre ratificamus approbamus ac pro nobis et successoribus nostris pro perpetuo confirmamus donationem infeofamentum et mortificationem per nostrum quondam charissimam matrem in sua perfecta etate factam datam et concessam predictis Preposito Balliuis Consulibus et Communitati dicti nostri Burgi et eorum successoribus pro ministrorum et pauperum infra idem supportatione et juvamine, de omnibus et singulis terris tenementis annuis redditibus alijsque proficuis et emolumentis quibuscunque jacentibus infra dictum nostrum Burgum et libertatem eiusdem que quouismodo perprius pertinuerunt ad quascunque capellanias collegia prebendaria fratres cuiuscunque ordinis aliasque personas beneficiatas, prout dicta donatio et mortificatio de data decimo tertio die mensis Martij anno Domini millesimo quingentesimo sexagesimo tertio sub magno sigillo dicte nostre quondam charissime matris latius proportat. Necnon aliam donationem et dispositionem prefatis Preposito Balliuis Consulibus et Communitati dicti nostri Burgi et eorum successoribus per nos nostro sub magno sigillo factam datam et concessam de Ecclesia Collegiata Trinitatis vulgo the Trinitie College nuncupata cum eiusdem cimiterio mansionibus domibus et hortis, cum hospitale Hospitalis

for us and our successors confirm for ever the gift, infeftment, and mortification made, given, and granted by our late dearest mother in her perfect age to the foresaid Provost, Bailies, Councillors, and Community of our said Burgh and their successors, for the support and help of the Ministers and Poor within the same, of all and sundry lands, tenements, annual rents, and other profits and emoluments whatsoever lying within our said Burgh and the liberty of the same, which formerly belonged in any manner of way to any chaplainaries, colleges, prebends, friars of whatsoever orders, and other beneficed persons, as the said gift and mortification, of date the thirteenth day of the month of March in the year of our Lord one thousand five hundred and sixty-three,[1] under the great seal of our said late dearest mother more fully sets forth. As also another gift and disposition given and granted by us under our great seal, to the said Provost, Bailies, Councillors, and Community of our said Burgh and their successors, of the Collegiate Church of the Trinity commonly called the Trinity College, with the cemetery, mansions, houses, and yards of the same, with the Hospital called the Hospital of Trinity College and yard of the same, so that the said Provost,

[1] Should be 1566.

Collegii Trinitatis nuncupato et horto eiusdem, sic ut dicti Prepositus Balliui et Consules vnum hospitale desuper construere et erigere possunt pro sustentatione pauperum honestorum seniorum et indigentium personarum infra dictum Burgum, prout dicta donatio et dispositio de data duodecimo die mensis Novembris anno Domini millesimo quingentesimo sexagesimo septimo similiter latius continet; Et etiam aliam donationem mortificationem et annexationem prepositure dicti Collegii Trinitatis cum omnibus terris redditibus proficuis et emolumentis, ac cum aduocatione et donatione oratorum pauperum vulgo lie beidmen et bedlaris dicti hospitalis, et omnibus aliis juribus et preuilegiis dicte prepositure spectantibus prout in dicta mortificatione de data vicesimo tertio die mensis Junii anno Domini millesimo quingentesimo octuagesimo quinto latius continetur: VNACUM nostra confirmatione et nova donatione dicte prepositure cum singulis terris proficuis et emolumentis eidem ac prebendariis et capellanis eiusdem collegii in proprietate seu communitate spectantibus, prout dicta donatio et noua dispositio de data vicesimo sexto die mensis Maii anno Domini millesimo quingentesimo octuagesimo septimo latius proportat. NEC NON annexa-

Bailies, and Councillors may construct and erect an Hospital thereupon for the sustentation of honest, poor, aged, and indigent persons within the said Burgh, as in the said gift and disposition, of date the twelfth day of the month of November in the year of our Lord one thousand five hundred and sixty-seven, is likewise more fully contained. As also another gift, mortification, and annexation of the provostry of the said College of Trinity, with all lands, rents, profits, and emoluments, and with the advocation and presentation of poor orators, in Scots called beidmen and bedlaris of the said Hospital, and all other rights and privileges belonging to the said provostry, as in the said mortification, of date the twenty-third day of the month of June in the year of our Lord one thousand five hundred and eighty-five, is more fully contained. Together with our confirmation and gift of new of the said provostry, with singular lands, profits, and emoluments belonging to the same, and to the prebendaries and chaplains of the same college, in particular or in common, as the said gift and disposition of new, of date the twenty-sixth day of the month of May in the year of our Lord one housand five hundred and eighty-seven, more fully sets forth. As also the an-

tionem archidiaconatus Laudonie cum terris redditibus decimisque garbalibus eidem spectantibus ad prefatum collegium infra dictum nostrum Burgum pro juventutis instructione nuper erectum annexatis et mortificatis prout hujusmodi de data quarto die mensis Aprilis anno Domini millesimo quingentesimo octuagesimo quarto proportat; Vnacum decreto dominorum nostri consilii et sessionis per quod decernitur et declaratur quod dicti Prepositus Ballivi Consules et Communitas jus ad decimas fructus emolumenta rectorie ecclesie de Dunbarny habent pro rationibus et causis in dicto decreto contentis, de data decimo nono die mensis Martii anno Domini millesimo quingentesimo octuagesimo tertio, in omnibus et singulis punctis capitibus et clausulis ac circumstantiis in eisdem particulariter et respective contentis. Insuper nos de nouo dedimus disposuimus et mortificauimus tenoreque presentis carte nostre damus concedimus disponimus et mortificamus prefatis Preposito Balliuis Consulibus et Communitati dicti nostri Burgi de Edinburgh eorumque successoribus pro ministrorum et pauperum sustentatione ac pro intertenemento dicti collegii per ipsos nuper erecti, omnes et singulas terras redditus decimas aliaque proficua et emolumenta de

nexation of the archdeaconry of Lothian, with the lands, rents, and teind sheaves belonging to the same, annexed and mortified to the foresaid College lately erected within our said Burgh for the instruction of youth, as the same, of date the fourth day of the month of April in the year of our Lord one thousand five hundred and eighty-four sets forth. Together with the decree of the Lords of our Council and Session, by which it is decerned and declared that the said Provost, Bailies, Councillors, and Community have right to the teinds, fruits, and emoluments of the parsonage of the church of Dunbarny for the reasons and causes contained in the said decree, of date the nineteenth day of the month of March in the year of our Lord one thousand five hundred and eighty-three, in all and singular the points, chapters, and clauses, and circumstances in the same particularly and respectively contained. Moreover, we of new have given, disponed and mortified, and by the tenor of this our present charter, give, grant, dispone, and mortify to the said Provost, Bailies, Councillors, and Community of our said Burgh of Edinburgh and their successors, for the support of the Ministers and Poor, and for the upholding of the said College by them lately erected, all and sundry lands,

particularibus in dictis superioribus donationibus et mortificationibus contenta ac in prefato decreto per dictos dominos consilii promulgato de data antedicta cum ipsis pro usibus in hujusmodi specificatis et contentis et non aliter, juxta formam et tenorem earundem in perpetuum remansura. Prouiso quod dicti Prepositus Balliui Consules et Communitas et eorum successores tenebuntur sustentare ministros apud suas ecclesias pro presenti ibidem seruientes, et similes qualificatas personas [qui] in hujusmodi curas in posterum ordinati erunt deseruire, secundum tenorem donationum et mortificationum dictis Preposito Balliuis Consulibus et Communitati eorumque successoribus perprius ad hunc effectum factarum vt premissum est. In cuius rei testimonium huic presenti carte nostre magnum sigillum nostrum apponi precepimus, Testibus etc. apud Halyrudehous vicesimo nono die mensis Julii anno Domini millesimo quingentesimo octuagesimo septimo, et regni nostri vicesimo primo.

rents, teinds, and other profits and emoluments, particularly contained in the said former gifts and mortifications, and in the foresaid decree pronounced by the said Lords of Council of the date aforesaid, to remain with them for ever for the uses therein specified and contained, and not otherwise, according to the form and tenor of the same. Providing that the said Provost, Bailies, Councillors, and Community, and their successors, shall be held bound to support the ministers in their churches serving there at present, and similar qualified persons who shall be ordained to serve in the same cures for ever, according to the tenor of the donations and mortifications to the said Provost, Bailies, Councillors, and Community, and their successors, formerly made to this effect as is premised. In witness whereof we have commanded our great seal to be affixed to this our present charter. Witnesses etc. At Holyroodhouse, the twenty-ninth day of the month of July in the year of our Lord one thousand five hundred and eighty-seven, and the twenty-first of our reign.

XVI.

CHARTER, commonly called "The Golden Charter," granted by King James VI., under his Great Seal, to the Provost, Bailies, Council, and Community of Edinburgh. Holyrood, 15th March 1603.

JACOBUS Dei gratia Rex Scotorum: Omnibus probis hominibus tocius terre sue clericis et laicis, salutem: SCIATIS quia nos post justam et legitimam viginti quinque annorum etatem, jam diu adimpletam, peractasque omnes tam generales quam speciales nostras revocationes; pro regia quam habemus sollicitudine considerantes insignem Civitatis nostre Edinburgene antiquitatem, quam illustrissimi felicissime memorie majores nostri in Regie Vrbis dignitatem et eminentiam erexerunt et sublimarunt, tum quod civium in exteras regiones navigantium, et negotiationem exercentium cura et industria, et indefessis laboribus Regii patrimonii nostri census adauctus sit, regnum locupletatum, et subditi nostri eorum exemplo ad humaniorem cultum traducti; eaque urbs in publicis oneribus subeundis, vectigalibus et tributis pendendis, exactionibus ceterisque impositionibus prestandis maximum pre ceteris

JAMES, by the grace of God king of Scots: To all good men of his whole land, clerics and laics, greeting: Know ye that we, long after we had completed the just and lawful age of twenty-five years, and had accomplished all our revocations as well general as special: considering for the royal solicitude which we have, the remarkable antiquity of our City of Edinburgh, which our most illustrious progenitors of most happy memory advanced and erected into the dignity and eminence of a royal City, and that through the care, industry, and indefatigable labours of the citizens sailing to foreign regions and exercising trade, the revenue of our royal patrimony has been increased, the kingdom enriched, and our subjects by their example led to more refined way of living, and that the said City in bearing public burdens, in paying taxes and tributes, and in furnishing exactions and other imposts, is subject to and sustains the greatest burden com-

urbibus onus impositum sentiat et sustinet, que non hac tempestate modo sed et multis retro seculis primaria princeps et totius regni maxime memorabilis, et ad patrie dignitatem illustrandam, et gloriam amplificandam aptissima extiterit, vtpote commodissimus publicis comitiis celebrandis locus, vbi nos et consiliarii, principes et aulici nostri, majore cum frequentia versamur, et solennes omnium conventus indici consueverunt; adeoque supremus senatus ciuilibus controuersiis judicandis constitutus, ab ipsis institutionis primordiis, et summus capitalium seu criminalium causarum judex, stationem et continuam sedem delegerunt. Ad hec memoria repetentes et certissime intelligentes, strenuam fidelem et acceptam operam, egregia facinora, beneficia memorabilia, seruitia et obsequia officii plena, ab hac inclyta vrbe ejusque civibus prestita, et per omnem oblatam occasionem tam pacis quam belli temporibus, tum nobis ipsis, tum illustrissimis predecessoribus nostris impensa, non tantum in regno hocce contra hostes externos propugnando, verum etiam corporum suorum objectu et facultatum dispendio, jura Majestatemque Regiam contra rebellium subditorum insultus protegendo, qui ciuilium discordiarum, internorum motuum, et intestinorum bellorum temporibus, aduersus regale diadema,

pared with other cities, and that it not only now is, but for many ages past has been, the first principal and most distinguished of the whole kingdom, and most fitted to shew forth the dignity and increase the fame of our country, as being the place most commodious for holding the public assemblies, where we, our councillors, nobles, and courtiers most frequently reside, and the stated general conventions are usually appointed, which also the supreme court constituted for the trial of civil suits, from its first institution, and the supreme judge of capital or criminal causes have chosen as their fixed quarters and seat: Moreover, calling to mind and perfectly understanding the strenuous, faithful, and acceptable endeavours, illustrious deeds, memorable benefits, services, and performances full of dutifulness, rendered by this renowned City and its citizens, and expended as well upon ourselves as on our most illustrious predecessors, on all occasions that offered in times both of peace and war, not only by fighting against foreign enemies in this kingdom, but also by the exposure of their bodies and the expenditure of their substance in defending our royal rights and majesty against the attack of rebellious subjects, who in times of civil discords, internal

partim pubescentibus, partim ad majorem etatem prouectis principibus, aliquid moliebantur cum cruenta sanguinis sui effusione, vite ciuium jactura, urbe sepius incendio absumpta, diuturna et summa solitudine, vastatione et facultatum direptione, in debita obsequii erga celsitudinem et Majestatem Regiam perseuerentia, majora detrimenta et damna equo animo ferendo, quam cetere pene uniuerse regni nostri vrbes sustinere potuerunt. Quemadmodum harum rerum monumenta in nonnullis vrbis innestituris et regiarum donationum literis, ad perpetuam eorum laudis et fame in pōsteros propagationem transmissa reperiuntur, et nos ipsi post regalem a nobis apicem susceptum non obscura eorum fidelitatis et intimi erga nos amoris et spontanee obedientie documenta multoties experti sumus atque etiamnum in dies experimur; cognoscentes preterea hujus nostre urbis magistratus et cives instaurationem et extensionem pire et portus sui Lethensis, constructionem fori et idonee apteque aree ingentibus trabibus undiquaque septe ad continenda et venumdanda ligna, tigna, et omnis generis materiam ligneam edificiorum usui maxime necessariam, refectionem platearum, structuram plurium sacrarum edium et templorum, et eorum cum conveniente pastorum

commotions, and intestine wars, attempted anything against the royal crown, sometimes when the princes were infants, sometimes when they had attained a greater age, with cruel effusion of their blood, loss of the lives of the citizens, the City itself often consumed by fire, with long-continued and extreme desolation, the wasting and plundering of their goods, in due perseverance in their duty towards the King's Highness and Majesty, bearing with equanimity greater damage and loss than almost all the other cities of our kingdom together were able to sustain. As memorials of these things are found preserved for the perpetuation of their praise and fame to posterity, in divers of their infeftments, and in letters of royal gifts: And we ourselves since we received the royal dignity, have many times experienced, and still daily experience, no obscure proof of their fidelity, hearty affection, and willing obedience to us: Understanding moreover, that the magistrates and citizens of this our City having not long ago undertaken the renovation and extension of their pier and port of Leith, the construction of a market place, and of a fit and proper area enclosed on all sides with a wooden fence for receiving and selling wood, logs, and timber of all sorts most needful for the erection of buildings, the repairing of streets, the

numero plantationem, datis ex publico stipendiis non levibus, erectionem et fundationem academiarum, ad literarum studia promovenda, hospitaliorum et tochodochiorum cum sufficienti redditu structuram et suppeditationem, aliaque magna et preclara opera publica, non ita pridem aggressos ad divini numinis gloriam, religionis et pietatis incrementum, et tocius regni nostri utilitatem, magna tum sedulitate promovisse, et majora quotidie moliri: In quibus perficiendis, ac majestati nostre maximi momenti negotiis per ingentium summarum numerationem adjuvandis, publici urbis census et redditus expensi, et privatorum civium patrimonia attenuata [et] facultates attrite sunt. Nos neque nulla in re majoribus nostris impares, neque munificentia liberalitate nec grati animi significatione erga eos omnes ditioni nostre subjectos, qui industria virtute et magnanimitate sua, quicquam de nobis promereri possunt inferiores, decrevimus huic nostre ciuitati ejusque ciuibus ob eorum sincerum erga nos amorem et fidelitatem, et obsequia et facta premiis et commendatione dignissima, perpetuum posteris monumentum et gratitudinis et fauoris nostri regii testimonium futuris seculis duraturum relinquere, non solum priscarum immunitatum et privilegiorum ratihabitione, inuestiturarum publicorum agrorum

building of more kirks and churches, and the planting of them with a sufficient number of ministers, liberal stipends out of the public funds being given to them, the erecting and founding of colleges for promoting the study of letters, the building of hospitals and houses for orphans, and supplying them with sufficient revenues, and other great and distinguished public works for the glory of the Divine name, the increase of religion and piety, and the advantage of our whole kingdom; promoted by them at that time with great diligence, and are daily contriving to do greater things: In completing which, and helping on affairs of the greatest moment to our Majesty, by the paying down of large sums of money, the common good and revenue of the city were expended, and the patrimonies of private citizens diminished and their means impaired: And we being neither in any wise unlike our predecessors, nor inferior to them in munificence, liberality, and the exhibition of gratitude towards all our subjects who by their industry, virtue, and magnanimity can at all deserve well of us, have resolved to leave to this our City and its citizens, for their sincere love and fidelity towards us, and services and deeds most worthy of reward and commendation, a perpetual monu-

censuum et regiarum largitionum approbatione, verum etiam nouorum majorum et digniorum preter supraque ceteras regni nostri vrbes et civitates amplificatione et augmentatione. His et aliis quam plurimis causis et rationibus equissimis nos moventibus et impellentibus, post nostram perfectam etatem, et diu post omnes revocationes nostras predictas, ex certa scientia et proprio motu, cum auisamento et consensu dominorum nostri secreti consilii, ac etiam cum expressis avisamento consensu et assensu nostrorum fidelium et familiarium consiliariorum domini Georgii Home de Spott militis nostri thesaurarii, ac domini Davidis Murray de Gosperdy militis nostrorum rotulorum computatoris, et Magistri Joannis Prestoun de Fentounbarnis nostri generalis collectoris ac thesaurarii nostrarum novarum augmentationum terrarum ecclesiasticarum hujus nostri regni ad nostram coronam, ratificavimus approbavimus ac pro nobis et nostris successoribus pro perpetuo confirmavimus, tenoreque presentis carte nostre ratificamus approbamus ac pro nobis et nostris successoribus pro perpetuo confirmamus omnes et quascunque cartas et infeofamenta, precepta, sasinarum instrumenta, confirmationes, acta, sententias, decreta, jura, titulos, securitates, literas, scripta, euidentias, donationes,

ment to posterity, and a testimony of our royal gratitude and favour that shall last to future generations, not only by ratifying their original immunities and privileges, confirming their infeftments in public lands, revenues, and royal gifts, but also by the enlargement and augmentation of new, greater, and more worthy [privileges] beyond and above the other cities and towns of our kingdom: For these and very many other most just causes and reasons us moving and impelling, after our perfect age, and long after all our foresaid revocations, of our certain knowledge and proper motion, with the advice and consent of the lords of our Privy Council, as also with the express advice, consent, and assent of our trusty and familiar councillors Sir George Home of Spott, knight, our treasurer, and Sir David Murray of Gosperdy, knight, our comptroller, and Master John Preston of Fentonbarns, our collector general and treasurer of our new augmentations of the church lands of this our kingdom to our crown, we have ratified, approved, and for us and our successors confirmed for ever, and by the tenor of this our present charter ratify, approve, and for us and our successors for ever confirm, all and whatsoever charters and infeftments, precepts, instruments of sasine, con-

concessiones, libertates, commoditates, immunitates et privilegia in eisdem contentas, factas et concessas seu confirmatas per nos nostrosque nobilissimos predecessores Reges et Reginas hujus nostri regni eorumque Gubernatores, et pro tempore Regentes, predicto nostro Burgo de Edinburgh, Prepositis, Aldermannis, Balliuis, Decanis Gilde, Thesaurariis, Consulibus, Burgensibus et Communitatibus ejusdem eorumque successoribus, ac ecclesiis collegiis ministris et hospitalibus dicti nostri Burgi, ex quacunque forma seu formis, contento seu contentis, data seu datis, sint; Et presertim minime generalitatem antedictam prejudicandi particulares cartas, infeofamenta, confirmationes, scripta et evidentias subtus specificatas, donationes, concessiones, libertates, immunitates et privilegia in eisdem contenta, videlicet.

. . . .

Cartam factam datam et concessam per nostram quondam charissimam matrem, de data decimo tertio die mensis Martii anno Domini millesimo quingentesimo sexagesimo sexto, regnique sui anno vigesimo quinto, dictis Preposito, Ballivis, Consulibus et Communitati dicti nostri Burgi de Edinburgh eorumque successoribus, de omnibus et singulis terris, tenementis, domibus, edificiis, ecclesiis, capellis, hortis, pomariis, croftis,

firmations, acts, sentences, decrees, rights, titles, securities, letters, writs, evidents, gifts, grants, liberties, commodities, immunities, and privileges therein contained, made and granted or confirmed by us and our most noble predecessors, the Kings and Queens of this our kingdom, and their Governors or Regents for the time, to our said Burgh of Edinburgh, the Provosts, Aldermen, Bailies, Deans of Guild, Treasurers, Councillors, Burgesses, and Communities thereof and their successors, and to the churches, colleges, ministers, and hospitals of our said Burgh, of whatever form or forms, content or contents, date or dates they be, and specially and without prejudice to the foresaid generality, the particular charters, infeftments, confirmations, writs and evidents after specified, gifts, grants, liberties, immunities, and privileges therein contained, videlicet:—

. . . .

A charter made, given, and granted by our late dearest mother, of date the thirteenth day of the month of March, in the year of our Lord one thousand five hundred and sixty-six, and of her reign the twenty-fifth, to the said Provost, Bailies, Councillors, and Community of our said Burgh of Edinburgh and their successors, of all and sundry lands, tenements, houses, buildings, churches, chapels,

annuis redditibus, decimis, fructibus, devoriis, proficuis, emolumentis, firmis lie almos dailsilver obeittis et anniversariis, que quovismodo pertinuerunt prius ad quascunque capellanias alteragia seu prebendarias in quascunque ecclesias capellas seu collegia, infra libertatem dicti nostri Burgi fundata per quoscunque patronos, quorum dicti capellani seu prebendarii in possessione fuerunt, ubicunque infra nostrum regnum aut infra seu extra nostrum Burgum de Edinburgh jacent; ac de omnibus et singulis terris, perprius ad Fratres Predicatores et Carmelitanos spectantibus, cum diversis aliis ad longum in dicta carta contentis.

. . . .

Cartam Confirmationis per nos concessam, antedictam cartam confirmantem de data decimo quarto die mensis Aprilis anno Domini millesimo quingentesimo octuagesimo secundo, regnique nostri anno decimo quinto. Cartam per nos factam et concessam de data vigesimo tertio die mensis Junii anno Domini millesimo quingentesimo octuagesimo quinto, regnique nostri anno decimo octavo, dictis Preposito Ballivis et Consulibus dicti nostri Burgi eorumque successoribus, de toto et integro beneficio prepositure Ecclesie Trinitatis Collegii prope predictum nostrum Burgum

yards, orchards, crofts, annual rents, teinds, fruits, duties, profits, emoluments, rents, alms, dailsilver, obits and anniversaries which in anywise belonged of-before to any chaplainries, altarages, or prebendaries founded in any churches, chapels, or colleges, within the liberty of our said Burgh, by whatsoever patrons, in possession whereof the said chaplains or prebendaries were, wherever they lie within our kingdom, either within or without our Burgh of Edinburgh; and of all and sundry lands formerly belonging to the Preaching Friars and Carmelite Friars, with sundry others contained at length in the said charter.

. . . .

A charter of confirmation granted by us confirming the foresaid charter of date the fourteenth day of the month of April, in the year of our Lord One thousand five hundred and eighty-two, and of our reign the fifteenth year. A charter made and granted by us of date the twenty-third day of the month of June, in the year of our Lord one thousand five hundred and eighty-five, and of our reign the eighteenth year, to the said Provost, Bailies, and Council of our said Burgh, and their successors, of all and whole the benefice of the provostry of the church of Trinity College, near our said Burgh of Edinburgh, with all

de Edinburgh, cum omnibus predictis terris ecclesiis decimis ac aliis ad idem spectantibus, ac cum ecclesiis parochialibus de Sowtray et Lempetlaw decimis et redditibus carundem ab antiquo ad dictam preposituram annexatis. Cartam Confirmationis per nos concessam, antedictam cartam confirmantem unacum nova donatione in eadem contenta de dicta prepositura, ac etiam de omnibus et singulis ecclesiis decimis fructibus et redditibus omnium partium et pertinentium dicte prepositure; necnon de omnibus fructibus et redditibus ad omnes et singulos prebendarios et capellanos dicti Collegii spectantibus et pertinentibus seu ad integros prebendarios in communia vel ad eorum aliquem in proprietate, de data vigesimo sexto die mensis Maii anno Domini millesimo quingentesimo octuagesimo septimo, regnique nostri anno vigesimo. Cartam per nos factam et concessam, de data duodecimo die mensis Novembris anno Domini millesimo quingentesimo sexagesimo septimo, prefatis Preposito Ballivis Consulibus et Communitati dicti nostri Burgi de Edinburgh eorumque successoribus, de Trinitatis Collegio, cum cemiterio ejusdem, mansionibus domibus et hortis, ac cum Hospitale nuncupato The Hospitall of the Trinitie College, et hortis ejusdem cum pertinentiis.

. . . .

the foresaid lands, churches, teinds, and others belonging to the same, and with the parish churches of Soltray and Lempetlaw, teinds and rents of the same annexed of old to the said provostry. A charter of confirmation granted by us confirming the foresaid charter, together with a new gift contained in the same, of the said provostry, as also of all and sundry churches, teinds, fruits, and rents, and of all parts and pertinents of the said provostry; as also of all fruits and rents belonging and pertaining to all and singular the prebendaries and chaplains of the said college, either to the whole of the prebendaries in common, or to any one of them in particular, of date the twenty-sixth day of the month of May, in the year of our Lord one thousand five hundred and eighty-seven, and of our reign the twentieth year. A charter made and granted by us of date the twelfth day of the month of November, in the year of our Lord one thousand five hundred and sixty-seven, to the foresaid Provost, Bailies, Councillors, and Community of our said Burgh of Edinburgh and their successors, of Trinity College with the cemetery of the same, mansions, houses, and yards, and with the hospital called The Hospitall of the Trinitie College, and yards of the same with the pertinents.

. . . .

UNACUM omnibus et singulis aliis cartis, infeofamentis, concessionibus, donationibus, privilegiis, immunitatibus, et juribus, parliamentorum, generalis conventionis, secretique consilii actis, sententiis et decretis per nos nostrosque nobilissimos progenitores, seu per quamcunque aliam personam seu personas, factis et concessis, ad et in favorem Aldermani, Prepositi, Ballivorum, Consiliariorum et Communitatum dicti nostri Burgi de Edinburgh pro tempore, eorumque predecessorum et successorum quorumcunque, erga et concernentibus erectionem antedicti nostri Burgi de Edinburgh in unum liberum Burgum regale, cum omnibus juribus, titulis, privilegiis ad idem spectantibus per leges et consuetudinem nostri regni, ac de omnibus terris, viis, plateis, meniis, muris, passagiis, moris, censubus, devoriis, lacubus, domibus, tenementis, hortis, dominiis, possessionibus, annuis redditibus, molendinis, terris molendinariis, multuris, amnibus, vulgo. lie dammys ripis precipiciis et omnibus earundem pertinentiis. . . . . . . ac de omnibus aliis libertatibus, redditibus, terris et jurisdictionibus, quas dicti Aldermani, Prepositi, Ballivi, Conciliarii et Communitas dicti nostri Burgi per seipsos eorumve predecessores, ullo tempore preterito, possidebant seu utebantur, aut quas pro presenti quovismodo utuntur et possident.

TOGETHER with all and sundry other charters, infeftments, grants, gifts, privileges, immunities and rights, acts of Parliament, of General Convention, and of Privy Council, sentences and decrees made and granted by us and by our most noble progenitors, or by any other person or persons, to and in favour of the Alderman, Provost, Bailies, Councillors, and Community of our said Burgh of Edinburgh for the time being, and their predecessors and successors whomsoever of and concerning the erection of our said Burgh of Edinburgh into a free Royal Burgh with all rights, titles, privileges belonging to the same by the laws and custom of our kingdom, and of all lands, ways, streets, bulwarks, walls, passages, muirs, rents, duties, lochs, houses, tenements, yards, lordships, possessions, annual rents, mills, mill lands, multures, dams, banks, haughs, and all the pertinents of the same. . . . . . . and of all other liberties, rents, lands, and jurisdictions which the said Aldermen, Provosts, Bailies, Councillors, and Community of our said Burgh by themselves or their predecessors possessed, or used in any time past, or which they at present use, and possess in any manner of way.

NECNON cum omnibus et singulis mortificationibus, patronatuum juribus, infeofamentis, donationibus et dispositionibus per nos et nostros predecessores, seu per aliquas alias personas spirituales seu seculares, factis et concessis, dictis Aldermano Preposito Ballivis Consulibus et Communitati dicti nostri Burgi, ministris, hospitalio et pauperibus ejusdem, eorumque predecessoribus, de omnibus et quibuscunque terris tenementis domibus edificiis hortis pomariis ecclesiis capellis capellanariis patronagiis collegiis annuis redditibus feudifirme firmis lie obeittis anniversariis lie dailsilver decimis tam rectoriarum quam vicariarum, ubicunque infra seu extra predictum nostrum Burgum jacent particulariter seu generaliter in eorum dispositionibus mortificationibus seu eorum aliquo contentis secundum formam et tenorem earundem; ac cum omnibus Parliamentorum seu generalis Consilii actis et aliis actis sententiis et decretis earundem aliquam partem seu easdem concernentibus. Ac volumus et concedimus, ac pro nobis et nostris successoribus cum avisamento predicto pro perpetuo decernimus et ordinamus, quod Antedicta generalitas specialitati minime damno seu prejudicio fuerit, et quod specialitas generalitati nullatenus derogaverit eamve prejudicaverit; et quod hec presens nostra confirmatio et premissorum

ALSO with all and sundry mortifications, rights of patronage, infeftments, grants and dispositions made and granted by us and our predecessors, or by any other persons, ecclesiastical or secular, to the said Alderman, Provost, Bailies, Councillors and Community of our said Burgh, the ministers, hospital, and poor of the same, and their predecessors, of all and whatsoever lands, tenements, houses, buildings, yards, orchards, churches, chapels, chaplainries, patronages, colleges, annual rents, feufarm rents, obits, anniversaries, daill silver, teinds as well parsonage as vicarage, wherever the same lie within or without our foresaid Burgh, as particularly or generally contained in their dispositions and mortifications, or any of them, after the form and tenor of the same: and with all acts of Parliament, or General Council, and other acts, sentences, and decrees concerning the same, or any part thereof: And we will and grant, and for us and our successors with advice foresaid for ever decern and ordain, that the aforesaid generality shall not in the least degree hurt or prejudice the speciality, and that the speciality shall not in any degree derogate from or prejudice the generality; and that this our present confirmation and approbation of the premises

approbatio sit et omnibus temporibus futuris fuerit tanti valoris roboris et in se efficacie et effectus in omnibus respectibus dicto Burgo nostro de Edinburgh, ac Preposito Ballivis Consulibus Burgensibus et Communitati ejusdem, eorumque successoribus, collegiis, ministris hospitalio ejusdemque pauperibus, ac si omnia et singula antedicta infeofamenta concessiones dispositiones mortificationes confirmationes, jura tituli securitates litere scripta et evidentie acta decreta et sententie omnesque donationes dispositiones libertates commoditates, immunitates et privilegia specialiter sive generaliter in eisdem mentionata ad longum de verbo in verbum hic insererentur; non obstante quod ratione multitudinis numerositatis longitudinis et eorundem prolixitatis minime hic inserta sint: super quibus nos pro nobis et successoribus nostris dispensavimus ac per presentis carte nostre tenorem dispensamus, pro nunc et imperpetuum. Insuper nos absque damno derogatione seu prejudicio antedictarum priorum cartarum, infeofamentorum, mortificationum, confirmationum, jurium, titulorum, securitatum, literarum, scriptorum et evidentiarum, actorum in Parliamentis, conventionibus et in secretis consiliis confectorum, sententiarum, actorum, decretorum, donationum, concessionum, libertatum, commoditatum,

and in all time coming shall be of as much force, strength, and effect in itself, in all respects to our said Burgh of Edinburgh, and to the Provost, Bailies, Councillors, Burgesses, and Community of the same, and their successors, [and to the] colleges, ministers, hospitals, and poor of the same, as if all and sundry the foresaid infeftments, grants, dispositions, mortifications, confirmations, rights, titles, securities, letters, writs, and evidents, acts, decrees, and sentences, and all gifts, grants, liberties, commodities, immunities, and privileges, specially or generally mentioned in the same, had been herein inserted at length word for word, notwithstanding that, by reason of the multitude, number, length, and prolixity of the same, they have not been here inserted, whereanent we, for us and our successors have dispensed, and by the tenor of this our present charter dispense now and for ever. Moreover we without hurt, derogation, or prejudice to the aforesaid former charters, infeftments, mortifications, confirmations, rights, titles, securities, letters, writings, and evidents, acts of Parliament, Convention, and Privy Council, sentences, acts, decrees, gifts, grants, liberties, commodities, immunities, and privileges, and in farther

immunitatum et privilegiorum,' ac in earundem majore corroboratione, accumulando jura juribus, DE NOVO cum avisamento et consensu antedictis pro bono fideli et gratuito servitio nobis nostrisque nobilissimis progenitoribus, per dictos Prepositum, Ballivos, Consiliarios, et Communitatem dicti nostri Burgi de Edinburgh, eorumque predecessores, prestito et impenso; ac ut ipsis in dicto servitio perseverandi meliorem occasionem prebeamus, dedimus, concessimus, assedavimus, arrendavimus, locavimus, et ad feudifirmam seu emphiteosim hereditarie dimisimus, et hac presenti carta nostra confirmavimus, tenoreque presentium cum avisamento predicto, damus, concedimus, assedamus, arrendamus, locamus, et ad feudifirmam seu emphiteosim hereditarie dimittimus, et hac presenti carta nostra confirmamus, prefatis predilectis nostris et fidelibus servitoribus, Preposito, Ballivis, Consulibus, Burgensibus et Communitati dicti Burgi nostri de Edinburgh, eorumque successoribus pro perpetuo, totum et integrum antedictum Burgum de Edinburgh, cum meniis, muris, fossis, portis, viis, plateis, passagiis, viis, stratis, terris, territoriis et communitate ejusdem, molendinis, terris molendinariis, multuris, amnis lie dammis, ripis, precipiciis, partibus et pertinentiis, antedictum Burgum in unum liberum Burgum Regale creando erigendo et constituendo cum omnibus et singulis libertatibus, privi-

corroboration thereof, *accumulando jura juribus*, of new, with advice and consent foresaid, for the good, faithful, and gratuitous service performed and rendered to us and our most noble progenitors, by the said Provost, Bailies, Councillors, and Community of our said Burgh of Edinburgh and their predecessors, and that they may have the more occasion to persevere in the said service, have given, granted, set, let, leased and in feu-farm heritably demitted, and by this our present charter have confirmed, and by the tenor of these presents, with advice foresaid, give, grant, let, rent, lease, and in feu-farm heritably demit, and by this our present charter confirm, to our foresaid well-beloved servants, the Provost, Bailies, Councillors, Burgesses, and Community of our said Burgh of Edinburgh, and their successors for ever, ALL and WHOLE the foresaid Burgh of Edinburgh with walls, bulwarks, ditches, gates, ways, streets, passages, causeways, lands, territories, and community of the same, mills, mill-lands, multures, dams, banks, haughs, parts and pertinents, creating, erecting, and constituting the foresaid Burgh into a free Royal Burgh, with all and

legiis, immunitatibus et jurisdictionibus, que per leges et consuetudinem nostri regni ad unum liberum Burgum Regale pertinuerunt, pertinent seu juste spectare poterint, infra citraque dictum nostrum Burgum, ac in et per omnes bondas in quantum viceecomitatus principalis de Edinburgh extenditur, seu in latitudine et longitudine tam regalitate quam regali, extendi poterit; ac specialiter a ripa vulgo nuncupata Edge Buklingbray ex orientali, ad aquam nuncupatam Almound Watter ex occidentali, et in quantum antedictus viceecomitatus extenditur versus austrum, ac ad medium aque, vulgo to the myd watter of Forth, ad boream.

. . . . .

PRETEREA nos cum avisamento et consensu predictis ex certa scientia et proprio motu univimus annexavimus et incorporavimus, tenoreque presentis carte nostre pro nobis et successoribus nostris unimus, annexamus et incorporamus ad et cum dicto nostro Burgo de Edinburgh, statu, privilegiis et libertatibus predictis eidem concessis, omnes et singulas predictas moras communes etc.

. . . .

sundry liberties, privileges, immunities, and jurisdictions which by the laws and custom of our kingdom have pertained, pertain, or can justly belong to a free Royal Burgh, in and within our said Burgh, and in and through all the bounds as far as the sheriffdom principal of Edinburgh extends, or may be extended, both in length and breadth, as well regality as royalty; and specially from the brae commonly called "Edge Buklingbray" on the east, to the water called "Almound Watter" on the west, and as far as the foresaid sheriffdom extends towards the south, and "to the myd watter of Forth" to the north.

. . . .

MOREOVER we, with advice and consent foresaid, of our certain knowledge and proper motion, have united, annexed, and incorporated, and by the tenor of our present charter, for us and our successors unite, annex, and incorporate to and with our said Burgh of Edinburgh, the estate, privileges, and liberties foresaid granted to the same, ALL and SUNDRY the foresaid common muir etc.

Ac etiam cum omnibus et singulis predictis terris ecclesiasticis annuis redditibus aliisque redditibus decimis et aliis supra recitatis infra et extra predictum nostrum Burgum, unacum dicta prepositura et prebendariis antedictis Trinitatis Collegii et hospitalis ejusdem, ac cum dicto Archidiaconatu Lowthonie, rectoria et vicaria de Currie, decimis et redditibus ejusdem, et cum dicto Collegio nuper infra predictum nostrum Burgum fundato, ac cum omnibus terris, ecclesiis decimis et redditibus ad eundem dotatis et annexatis, cum pertinentiis, in unum liberum Burgum Regale tenementum et tenandriam; ac volumus et concedimus ac pro nobis nostrisque successoribus pro perpetuo decernimus et ordinamus quod dicti Prepositus, Ballivi, Decanus Gilde, Thesaurarius, Consules, Communitas, et Burgenses dicti nostri Burgi de Edinburgh, eorumque successores omnibus temporibus futuris, libere et pacifice gaudebunt et possidebunt idem Burgum nostrum de Edinburgh, cum communibus moris, etc.

. . . .

Cum integris antedictis terris ecclesiasticis beneficiis decimis redditibus ac aliis particulariter et generaliter supra recitatis, cum omnibus earundem pertinentiis quibuscunque, hic pro brevitate minime repetitis, tanquam

As also with all and sundry the foresaid church lands, annual rents and other rents, teinds, and others above recited, within and without our foresaid Burgh, together with the said provostry and prebends of the foresaid Trinity College and Hospital of the same, and with the said Archdeaconry of Lothian, parsonage and vicarage of Curry, teinds and rents of the same, and with the said College lately founded within our said Burgh, and with all lands, churches, teinds, and rents given and annexed to the same with the pertinents, into a free Royal Burgh, tenement and tenandry; And we will and grant, and for us and our successors for ever decern and ordain that the said Provost, Bailies, Dean of Guild, Treasurer, Councillors, Community, and Burgesses of our said Burgh of Edinburgh and their successors in all time coming shall freely and peaceably enjoy and possess our said Burgh of Edinburgh, with the common muirs, etc.

. . . .

With the whole foresaid church lands, benefices, teinds, rents, and others particularly and generally above recited, with all pertinents of the same whatsoever, here for brevity not repeated, as a free tenement and tenandry, in feu farm and

unum liberum tenementum et tenandriam in feudifirma et libero burgagio imperpetuum: et quod unica sasina semel danda virtute hujus presentis infeofamenti modo et in forma subsequenti, Preposito seu alicui uni Ballivorum prefati nostri Burgi de Edinburgh pro tempore, apud crucem foralem dicti nostri Burgi, stabit et ipsis eorumque successoribus sufficiens sasina erit, pro antedicto Burgo ac pro integris annexis connexis incorporationibus ac aliis particulariter et generaliter supra recitatis nunc ad idem unitis et annexatis ut premissum est, non obstante quod antedictum Burgum terre portus vie platee passagia custume privilegia libertates jurisdictiones ecclesie decime redditus ac alia particulariter et generaliter suprarecitata ad idem spectantia non jacent insimul et contigue sed in diversis partibus, super quo nos pro nobis et successoribus nostris dispensavimus, ac per presentis carte nostre tenorem dispensamus pro nunc et imperpetuum. Et nos considerantes dictos Prepositum Ballivos Consules et Communitatem dicti nostri Burgi Universitatem et Communitatem esse que in sua natura permaneat nullum particularem successorem habentem, ideo declaramus volumus et concedimus ac pro nobis et successoribus nostris decernimus et ordinamus quod antedicta sasina semel capta

free burgage for ever. And that one sasine once given by virtue of this present infeftment in manner and in form following to the Provost or any one of the Bailies of our foresaid Burgh of Edinburgh for the time, at the market cross of our said Burgh, shall stand and be sufficient sasine to them and their successors for the foresaid Burgh, and for the whole annexis, connexis, incorporations, and others particularly and generally above recited, now united and annexed to the same as is premised, notwithstanding that the foresaid Burgh, lands, ports, ways, streets, passages, customs, privileges, liberties, jurisdictions, churches, teinds, rents, and others particularly and generally above recited belonging to the same, do not lie together and contiguous, but in divers parts, whereanent we for us and our successors have dispensed, and by the tenor of our present charter dispense for now and ever. And we, considering the said Provost, Bailies, Councillors, and Community of our said Burgh to be a Corporation and Community which in its own nature endures, having no particular successor, therefore we declare, will, and grant, and for us and our successors decern and ordain, that the foresaid sasine once taken, by virtue of this our present new

virtute hujus presentis nostri novi infeofamenti per Prepositum seu per unum Ballivorum dicti nostri Burgi nomine omnium Burgensium et Communitatis ejusdem eorumque successorum, per deliberationem fundi et lapidis pro antedicto Burgo, terris molendinis et aliis accessoribus et ejusdem dependentiis, ac per deliberationem unius aurei nummi pro censubus devoriis custumis ac aliis accessoribus et eorum dependentiis, et per deliberationem unius basti pro officiis et jurisdictionibus predictis ac superioritate antedicte ville de Leith ac aliis accessoribus et eorundem dependentiis; ac per deliberationem unius libri psalmorum pro antedictis beneficiis ecclesiis decimis redditibus ac aliis accessoribus et eorundem dependentiis stabit et pro perpetuo erit sufficiens, sine ulla renovatione reiteratione seu recuperatione nove alicujus sasine imposterum. TENENDUM ET HABENDUM totum et integrum predictum Burgum de Edinburgh.

. . . .

NECNON omnes et singulas predictas terras tenementa domos edificia ecclesias capellas hortos croftas annuos redditus fructus devorias proficua emolumenta firmas elimozinas, lie almosdail silver obeitis, et anniversarias ad quascunque capellanias alteragia seu prebendarias in quibuscunque

infeftment, by the Provost or by one of the Bailies of our said Burgh, in name of all the Burgesses and Community of the same and their successors, by delivery of earth and stone for the foresaid Burgh, lands, mills, and other accessories and dependencies of the same, and by the delivery of a golden penny for the rents, duties, customs, and others their accessories and dependencies, and by the delivery of a baton for the offices and jurisdictions foresaid, and for the superiority of the foresaid town of Leith, and others their accessories and dependencies, and by the delivery of a psalm book for the foresaid benefices, churches, teinds, rents, and others the accessories and dependencies of the same, shall stand and be for ever sufficient, without any renovation, retaking, or obtaining of a new sasine hereafter: To HAVE and TO HOLD all and whole the foresaid Burgh of Edinburgh, etc.

. . . .

As ALSO all and sundry the foresaid lands, tenements, houses, buildings, churches, chapels, yards, crofts, annual rents, fruits, duties, profits, emoluments, rents, alms, dail silver, obits, and anniversaries pertaining and belonging to whatsoever chaplainries, altarages, or prebends, belonging or pertaining to what-

ecclesiis capellis seu collegiis infra libertatem dicti nostri Burgi fundatas spectantia et pertinentia; ac omnes et singulas antedictas terras ad Fratres Predicatores et Carmelitanos dicti nostri Burgi spectantes cum omnibus earundem pertinentiis quibuscunque; totum et integrum antedictum beneficium prepositure antedicte ecclesie Trinitatis Collegii, prebendariarum et capellaniarum ejusdem, cum omnibus et singulis terris ecclesiis, decimis ac aliis ad idem in communia seu proprietate spectantibus et pertinentibus; Ac cum antedictis ecclesiis parochialibus de Sowtray et Lempetlaw decimis et redditibus earundem ab antiquo ad dictam preposituram annexatis; Ac cum predicto Trinitatis Collegio cemiterio mansionibus domibus et hortis ejusdem ac cum dicto Hospitali dicti Trinitatis Collegii, hortis et omnibus ejusdem pertinentibus, et totum et integrum predictum Archidiaconatum Lowthonie, rectoriam et vicariam de Curry, cum decimis fructibus redditibus mansione gleba et terris ecclesiasticis ejusdem . . . . . . prefatis Preposito, Ballivis, Decano Gilde, Thesaurario, Consulibus, Burgensibus et Communitati dicti nostri Burgi de Edinburgh eorumque successoribus, de nobis et successoribus nostris, in feodo hereditate et libero burgagio, ac liberis officiis vice-

soever churches, chapels, or colleges founded within the liberty of our said Burgh; and all and sundry the foresaid lands belonging to the Friars Preachers and Carmelites of our said Burgh, with all the pertinents of the same whatsoever; all and whole the foresaid benefice of the Provostry of the foresaid church of Trinity College, prebends and chaplainries of the same, with all and sundry lands, churches, teinds and others belonging and pertaining to the same in commonty or property; and with the foresaid parish churches of Soltray and Lempitlaw, teinds and rents of the same, annexed of old to the said provostry; and with the foresaid Trinity College, cemetery, mansions, houses, and yards of the same, and with the said Hospital of the said Trinity College, yards, and all the pertinents of the same, and all and whole the foresaid Archdeaconry of Lothian, parsonage, and vicarage of Curry, with the teinds, fruits, rents, manse, glebe, and church lands of the same . . . . . . to the said Provost, Bailies, Dean of Guild, Treasurer, Councillors, Burgesses, and Community of our said Burgh of Edinburgh and their successors, of us and our successors, in fee, heritage, and free burgage, and with the free offices of sheriff, justiciary, and

comitis justiciarii et coronatoris infra predictas boudas pro nunc et imperpetuum, per omnes rectas metas suas antiquas et divisas, prout jacent in longitudine et latitudine, in domibus edificiis hortis boscis planis moris marresiis viis semitis aquis stagnis rivolis pratis pascuis et pasturis molendinis multuris et eorum sequelis, aucupationibus venationibus piscationibus, petariis turbariis carbonibus carbonariis, cuniculis cunicolariis columbis columbariis, fabrilibus brasinis brueriis, et genestis silvis nemoribus et virgultis lignis tignis lapicidiis lapide et calce, cum curiis et earum exitibus, herezeldis bludewitis et mulierum merchetis, cum pit et gallous, infangthief et outfangthief, cum communi pastura liberoque introitu et exitu, ac cum omnibus et singulis libertatibus commoditatibus proficuis et asiamentis ac justis suis pertinentiis quibuscunque, tam non nominatis quam nominatis, tam subtus terra quam supra terram, procul et prope, ad predictum Burgum, terras molendina officia portus libertates privilegia superioritatem annuos redditibus beneficia decimas ac alia particulariter et generaliter respective supra specificata, cum suis pertinentiis spectantibus seu juste spectare valentibus quomodolibet in futurum, libere quiete plenarie integre honorifice bene et in pace absque ulla revocatione contradictione

coroner within the foresaid bounds now and for ever, by all their just ancient marches and divisions, as they lie in length and breadth, in houses, buildings, yards, woods, plains, muirs, marshes, ways, paths, waters, ponds, streams, meadows, pastures and pasturages, mills, multures and their sequels, hawkings, huntings, fishings, peats, turfs, coals, coalpits, rabbits, rabbit warrens, doves, dovecots, forges, malt kilns, breweries, and heaths, woods, groves, and thickets, wood, timber, quarries, stone and lime, with courts and their issues, herezelds, bloodwits and merchets of women, with pit and gallows, infangthief and outfangthief, with common pasturage and free ish and entry, and with all and sundry liberties, commodities, profits, and easements, and their just pertinents whatsoever, as well not named as named, as well under the earth as above the earth, far and near, belonging or that could justly belong in any manner of way in time coming to the foresaid Burgh, lands, mills, offices, port, liberties, privileges, superiority, annual rents, benefices, teinds, and others, particularly and generally respectively above specified, with their pertinents freely, quietly, fully, wholly, honourably, well, and in peace, without any

impedimento aut obstaculo quocunque. Reddendo annuatim dicti Prepositus Ballivi Consules Decanus Gilde Communitas Burgenses et Inhabitatores dicti nostri Burgi de Edinburgh, eorumque successores, nobis et successoribus nostris pro antedicto Burgo de Edinburgh, etc. . . . . . quinquaginta duas mercas sterlingas ad terminos Penthecostes et Sancti Martini in hieme pro equali portione, nomine feudiferme. . . . .

. . . .

Et etiam pro omnibus et singulis predictis terris ecclesiasticis beneficiis decimis annuis redditibus aliisque redditibus et dependentiis earundem, dicti Prepositus Ballivi Consules et Communitas dicti nostri Burgi, et ministri apud dictas ecclesias pauperesque antedictorum hospitalium, scolasticique dictorum collegiorum et scolarum, eorumque successores, faciendo quotidie devotas et humiles preces Deo Omnipotenti, pro preservatione nostri successorumque nostrorum, secundum formam et tenorem priorum infeofamentorum earundem respective tantum. In cujus rei testimonium huic presenti carte nostre, magnum sigillum nostrum apponi precipimus. Testibus, predilectis nostris consanguineis

revocation, contradiction, impediment, or obstacle whatsoever: Paying yearly the said Provost, Bailies, Councillors, Dean of Guild, Community, Burgesses, and Inhabitants of our said Burgh of Edinburgh and their successors to us and our successors for the foresaid Burgh of Edinburgh . . . . . fifty-two merks sterlings at the terms of Whitsunday and Martinmas by equal portions in name of feu farm.

. . . .

And also for all and sundry the foresaid church lands, benefices, teinds, annual rents, and other rents and dependencies of the same, the said Provost, Bailies, Councillors, and Community of our said Burgh, and the ministers at the said churches, and poor of the said hospitals, and scholars of the said colleges and schools and their successors making devout and humble prayers daily to Almighty God for the preservation of us and our successors according only to the form and tenor of the prior infeftments of the same respectively. In witness whereof we have commanded our great seal to be affixed to this our present charter. Witnesses, our well-beloved cousins and councillors, John

et consiliariis, Joanne marchione de Hammiltoun comite Arranie domino Evane, Joanne comite de Montrois domino Grahame, etc. cancellario nostro, Georgio comite Mariscalli domino Keyth, etc. regni nostri mariscallo; dilectis nostris familiaribus consiliariis, dominis Jacobo Elphinstoun de Barntoun, nostro secretario; Richardo Cokburne juniore de Ormistoun militibus, nostro justiciarie clerico; et Magistro Willelmo Scott de Elie, nostre cancellarie directore. Apud Halyrudehous decimo quinto die mensis Martii, anno Domini millesimo sexcentesimo tertio, regnique nostri anno tricesimo sexto.

marquis of Hamilton earl of Arran lord Evane, John earl of Montrose lord Graham, etc., our chancellor, George earl Marischall lord Keith, etc., marischall of our kingdom; our beloved familiar councillors, Sirs James Elphinstoun of Barntoun, our secretary, Richard Cokburn younger of Ormiston, knights, our justice-clerk; and Master William Scott of Elie, director of our chancery. At Holyrood House the fifteenth day of the month of March in the year of our Lord one thousand six hundred and three, and in the thirty-sixth year of our reign.

XVII.

CHARTER by King James the Sixth, under his Great Seal, confirming previous Grants of the Kirk-livings, and of new granting the same to the Provost, Bailies, Council, and Community of the Burgh of Edinburgh. Beauvoir Castle, 7th August 1612.

JACOBUS Dei gratia Magne Britannie Francie et Hibernie Rex fideique defensor: Omnibus probis hominibus totius terre sue clericis et laicis, salutem: Sciatis quia nos, cum avisamento et consensu dominorum nostri secreti consilii regni nostri Scotie, nostrorum commissionariorum, officiarios status ejusdem regni representantium, intelligentes quod nos et quondam nostra charissima mater Maria Dei gratia Regina Scotorum, aliique nostri preclarissimi progenitores bone memorie, virtute diversorum infeofamentorum mortificationum annexationum donationum et dispositionum in favorem Prepositi Ballivorum Consulum et Communitatis Burgi nostri de Edinburgh, eorumque successorum, factorum datorum et concessorum, pro incremento policie infra dictum Burgum, honesta sustentatione ministrorum verbi Dei curam apud ecclesias ejusdem Burgi inservientium, magistrorum regen-

JAMES, by the grace of God of Great Britain France and Ireland, King, and Defender of the Faith: To all good men of his whole land, clerics and laics, greeting: Know ye that we, with the advice and consent of the lords of our Privy Council of our kingdom of Scotland, our commissioners representing the officers of state of the same kingdom, understanding that we and our late dearest mother Mary, by the grace of God Queen of Scots, and others our most noble progenitors of worthy memory, by virtue of divers infeftments, mortifications, annexations, gifts and dispositions, made, given and granted in favour of the Provost, Bailies, Councillors and Community of our Burgh of Edinburgh and their successors, for the increase of policy within the said Burgh, honest sustentation of the ministers of God's word serving the cure at the churches of the same Burgh, the masters, regents and others professors of liberal sciences, and

tium aliorumque liberalium scientiarum professorum et curam infra Collegium dicti Burgi inseruientium, et pro sustentatione hospitalis pauperum mutilatorum et indigentium personarum orphanorum et infantum infra dictum Burgum parentibus orbatorum, dedimus concessimus et disposuimus prefatis Preposito Ballivis Consulibus et Communitati dicti Burgi nostri de Edinburgh, et eorum successoribus imperpetuum omnes et singulas terras tenementa domos edificia ecclesias capellas hortos pomaria croftas annuos redditus fructus devorias proficua emolumenta census eleemosinas lie dailsiluer obitus et anniversaria quecunque que quovismodo ad quascunque capellas alteragia et prebendas fundatas in quacunque ecclesia capella aut collegio infra libertatem dicti Burgi nostri de Edinburgh, per quoscunque patronos eorundem, pertinuerunt aut pertinere dignoscentur, in quorum possessione capellani et prebendarii eorundem per prius fuerunt, ubicunque dicta tenementa domus edificia horti pomaria annui redditus anniversaria fructus proventus et emolumenta jacent, aut antea levata respective fuerunt, cum maneriorum locis pomariis terris annuis redditibus emolumentis et devoriis quibuscunque, que per prius ad Fratres Dominicanos Predicatores et Minores seu Franciscanos dicti

serving the cure within the College of the said Burgh, and for the entertainment of the Hospital's poor, mutilated and indigent persons, orphans and infants destitute of parents, within the said Burgh, have given, granted and disponed to the foresaid Provost, Bailies, Councillors and Community of our said Burgh of Edinburgh and their successors for ever, All and sundry lands, tenements, houses, buildings, churches, chapels, yards, orchards, crofts, annual rents, fruits, duties, profits, emoluments, maills, alms, daillsilver, obits and anniversaries whatsoever, which in any way pertain or are known to pertain to whatsoever chaplainries, altarages and prebends founded in whatsoever church, chapel, or college within the liberty of our said Burgh of Edinburgh, by whatsoever patrons of the same, in possession whereof the chaplains and prebendaries of the same formerly were, wheresoever the said tenements, houses, buildings, yards, orchards, annual rents, anniversaries, fruits, provents and emoluments lie, or were formerly uplifted respectively, with the manor places, orchards, lands, annual rents, emoluments and duties whatsoever which formerly belonged to the Friars Dominican Preachers and Minorites or Franciscans of our said

Burgi nostri de Edinburgh pertinuerunt; unacum omnibus et singulis terris tenementis et domibus infra dictum Burgum et libertatem ejusdem jacentibus, cumque omnibus annuis redditibus de quibuscunque domibus terris seu tenementis infra dictum Burgum levandis, quibuscunque capellaniis alteragiis ecclesiis mortuariis seu anniversariis fundatis datis seu mortificatis, ubicunque eadem jacent infra dictum regnum nostrum Scotie: Et similiter cum omnibus et singulis annuis redditibus aliisque devoriis solitis et consuetis, aut que per quamcunque ecclesiam extra dictum Burgum a Preposito et Ballivis ejusdem e communibus redditibus dicti Burgi, pro suffragiis faciendis cum pertinentiis, peti seu acclamari poterant; nec non totum et integrum beneficium prepositure Collegii Trinitatis prope dictum Burgum situati, cum omnibus et singulis ecclesiis decimis garbalibus aliisque decimis mansis domibus edificiis hortis pomariis annuis redditibus advocationibus donationibus et juribus patronatuum prebendariarum et capellaniarum, ac donatione oratorum pauperum vulgo lie beidmen et bedlyaris nuncupatorum, aliorumque officiariorum dicte prepositure et Hospitalis Trinitatis Collegii, ejusdem spectantibus et incumbentibus; unacum ecclesiis parochialibus de Sowtray et Lempitlaw aliisque ecclesiis et decimis ad dictam preposi-

Burgh of Edinburgh; together with all and sundry lands, tenements and houses lying within the said Burgh and liberty of the same, and with all annual rents to be uplifted furth of whatsoever houses, lands, or tenements within the said Burgh, founded, given and doted to whatsoever chaplainries, altarages, churches, burials, or anniversaries, wheresoever the same lie within our said kingdom of Scotland; and likewise with all and sundry annual rents and other duties used and wont, or which may be asked or claimed by whatsoever church outwith the said Burgh from the Provost and Bailies of the same, furth of the common rents of the said Burgh, for making of suffrages, with the pertinents; also All and Whole the benefice of the provostry of the Trinity College situated near the said Burgh, with all and sundry churches, teind sheaves and other teinds, manses, houses, buildings, yards, orchards, annual rents, advocations, donations, and rights of patronage of prebendaries and chaplainries and presentation of poor orators, in Scots called beidmen and bedlyaris, and other officers of the said Provostry and Hospital of Trinity College, belonging to, and incumbents of the same; together with the parish churches of Soltray and Lempitlaw, and

turam annexatis, cumque loco pomario et horto nuncupato lie Dingwall Castell ad dictum beneficium spectante, omnibusque et singulis fructibus redditibus emolumentis juribus casualitatibus proficuis devoriis tenentibus tenandriis et justis pertinentiis ad dictam prepositurам pertinentibus, ubicunque infra dictum regnum nostrum Scotie eadem jacent, per dictos Prepositum Ballivos Consules eorumque successores, pro sustentatione ministrorum dicti Burgi, seniorum decrepitorum orphanorum et pauperum infra dicta hospitalia, et pauperum scholasticorum infra dictum Burgum et scholas ejusdem, in futurum intromittendis colligendis utendis et disponendis: Et similiter quum nobis dictisque nostri secreti consilii dominis innotescat nos perprius annexasse ad Collegium dicti Burgi acras terrarum locum et tenementa Ecclesie de Campis lie Kirk of Field, infra libertatem dicti nostri Burgi situate, unacum rectoria et vicaria ecclesie de Currie decimis garbalibus aliisque decimis fructibus emolumentis et devoriis quibuscunque eedem spectantibus et incumbentibus, manso gleba terris ecclesiasticis et earundem pertinentiis pro magistrorum regentium aliorumque professorum curam infra dictum Collegium inservientium, sustentatione, per dictos Prepositum Ballivos

other churches and teinds annexed to the said provostry, with the place, orchard and yard called Dingwall Castell belonging to the said benefice, and all and sundry fruits, rents, emoluments, rights, casualties, profits, duties, tenants, tenandries and just pertinents belonging to the said provostry, wherever the same lie within our said kingdom of Scotland, to be intromitted with, collected, used and disponed upon by the said Provost, Bailies, Councillors, and their successors, for the entertainment of the ministers of the said Burgh, of the aged decrepit, orphans and poor within the said hospitals, and of poor scholars within the said Burgh and schools of the same, in time coming; And in like manner since it has become known to us and to the said lords of our Privy Council that we had formerly annexed to the College of the said Burgh the acres of land, place and tenements of the Church of the Fields, or Kirk of Field, situated within the liberty of our said Burgh, together with the parsonage and vicarage of the church of Curry, teind sheaves and other teinds, fruits, emoluments and duties whatsoever pertaining and belonging thereto, manse, glebe, church lands and pertinents of the same, for sustentation of the masters, regents and others professors serving the cure within the said College, to be intromitted with, uplifted, used and disponed upon by

Consules et Communitatem dicti nostri Burgi, ad utilitatem dicti Collegii magistrorum regentium aliorumque professorum curam in eodem Collegio inservientium intromittendis levandis utendis et disponendis, prout infeofamentis mortificationibus annexationibus donationibus et dispositionibus respective, tam per nos quam nostram charissimam matrem aliosque nostros preclarissimos progenitores super premissis factis et concessis, latius continetur: Quemadmodum nos cum avisamento dicti regni nostri Scotie statuum in Parliamento congregatorum tento apud Edinburgum infra Pretorium ejusdem, quinto die mensis Junii anno Domini millesimo quingentesimo nonagesimo secundo, ratificavimus et approbavimus donationes et mortificationes factas per dictam quondam nostram charissimam matrem, terrarum beneficiorum et redditaum fundatorum et mortificatorum pro ministrorum infra dictum Burgum nostrum de Edinburgh et hospitalium ejusdem sustentatione, ac etiam de novo ad Communitatem dicti Burgi eorumque successores, in favorem dicti ministerii suorumque hospitalium, annexavimus omnia et singula predicta tenementa annuosque redditus infra libertatem dicti Burgi jacentes, beneficia fundata et admortizata, omnesque terras et annuos redditus [extra libertatem] ejusdem Burgi jacentes annexatas ad quod-

the said Provost, Bailies, Councillors and Community of our said Burgh for the use of the said College, masters, regents and other professors serving the cure in the same College, as in the infeftments, mortifications, annexations, gifts and dispositions respectively made and granted as well by us as by our dearest mother and others our most noble progenitors concerning the premises, is more fully contained. Likeas we, with the advice of the Estates of our kingdom of Scotland convened in the Parliament held at Edinburgh within the Tolbooth of the same on the fifth day of June in the year of our Lord one thousand five hundred and ninety-two, have ratified and approved the donations and mortifications made by our said late dearest mother, of the lands benefices and rents founded and mortified for the sustentation of the ministers within our said Burgh of Edinburgh and hospitals of the same; and also have of new annexed to the Community of the said Burgh, and their successors, in favour of the said ministry and their hospitals, all and sundry the foresaid tenements and annual rents lying within the freedom of the said Burgh, benefices founded and mortified, and all lands and annual rents lying [outwith the freedom] of the said Burgh,

cunque beneficium prebendam aut locum, religiosum situatas infra libertatem dicti Burgi, cum omnibus proficuis emolumentis feudifirme firmis censibus et devoriis hujusmodi, dictosque Prepositum Ballivos Consules et Communitatem eorumque successores surrogavimus in pleno jure et titulo omnium terrarum annuorum reddituum et emolumentorum infra libertatem dicti Burgi, que per prius ad quemcunque episcopum abbatem commendatarium priorem aut aliquam aliam personam ecclesiasticam infra dictum regnum nostrum Scotie, pertinuerunt, et novum infeofamentum pro eorum securitate, si expediens fuerit, desuper expedire ordinavimus; nec non ad effectum predictum, nos cum avisamento dictorum statuum nostri Parliamenti dissolvimus generalem annexationem in illa parte ejusdem, in quantum eadem extendi apparere poterit ad aliqua premissorum, sive ad annexationem perprius factam in favorem dicti Collegii et Hospitalis dicti nostri Burgi, ecclesie de Dumbarnie cujusquidem ecclesie de Dumbarnie ecclesie de Potie et Moncreif sunt pendicula, jacentes infra vicecomitatum nostrum de Perth, rectorie et vicarie de Currie cum decimis fructibus redditibus emolumentis proficuis et devoriis ad easdem spectantibus et pertinentibus, ac etiam terrarum annuorum reddituum domorum edificiorum et hortorum dicti

annexed to any benefice, prebend, or religious place situated within the freedom of the said Burgh, with all profits, emoluments, feu-fermes, maills and duties thereof; and we have surrogated the said Provost, Bailies, Councillors and Community and their successors in the full right and title of all the lands, annual rents, and emoluments within the freedom of the said Burgh, which formerly pertained to whatsoever bishop, abbot, commendator, prior, or any other ecclesiastical person within our kingdom of Scotland; and we have ordained a new infeftment to be expede thereupon for their security, if it shall be thought expedient. Also for the foresaid effect, we, with the advice of the said estates of our Parliament, have dissolved the general annexation in that part of the same, in so far as the same might appear to be extended to any of the premises, or to the annexation previously made in favour of the said College and Hospital of our said Burgh, of the church of Dumbarnie, of which church of Dumbarnie the churches of Potie and Moncreiff are pendicles, lying within our sheriffdom of Perth, of the parsonage and vicarage of Currie with the teinds, fruits, rents, emoluments, profits and duties belonging and pertaining to the same, as also of the lands,

Trinitatis Collegii infra dictum Burgum situati, tam ad prepositum quam ad prebendarios ejusdem spectantium, nec non communium terrarum et annuorum reddituum hujusmodi et integre vicarie dicte ecclesie de Dumbarnie cum decimis fructibus redditibus emolumentis proficuis et devoriis ad eandem spectantibus et pertinentibus, cum prefatis Preposito Ballivis Consulibus et Communitati dicti nostri Burgi eorumque successoribus, in futurum remansuris, pro dicti sui ministerii et hospitalis sustentatione; Ac etiam nos cum avisamento dictorum statuum nostri Parliamenti declaravimus quod prefati Prepositus Ballivi Consules et Communitas eorumque successores omni tempore futuro habuerunt habent et habebunt tale plenum jus proprietatis et superioritatis dictarum terrarum annuorum reddituum et revenuorum tenentium tenendriarum et libere tenentium servitiorum earundem, sicuti habuerunt episcopi abbates priores fratres monachi sorores moniales capellani et prebendarii ad quos dicte terre et annui redditus perprius pertinuerunt, non obstante aliquo acto aut constitutione dictum actum nostri Parliamenti precedente, prout in eodem latius continetur: Igitur nos cum avisamento et consensu dictorum dominorum nostri secreti consilii, nostrorum commissionariorum predictorum, nunc, post nostram

annual rents, houses, buildings, and yards of the said Trinity College situated within the said Burgh, pertaining as well to the provost as to the prebendaries of the same, and of the common lands and annual rents of the same, and of the whole vicarage of the said church of Dumbarnie, with teinds, fruits, rents, emoluments, profits and duties belonging and pertaining to the same, to remain with the foresaid Provost, Bailies, Councillors and Community of our said Burgh and their successors in time coming, for the sustentation of their said ministry and hospital. And also we, with the advice of the said estates of our Parliament, have declared that the foresaid Provost, Bailies, Councillors and Community, and their successors, in all time coming, had, have, and shall have as full right of property and superiority of the said lands, annual rents and revenues, tenants, tenandries and service of free tenants of the same, as had the bishops, abbots, priors, friars, monks, sisters, nuns, chaplains and prebendaries to whom the said lands and annual rents formerly belonged, notwithstanding any act or constitution preceding the said Act of our Parliament, as in the same is more fully contained. Therefore we, with the advice and consent of the said

legitimam et perfectam etatem omnesque nostras revocationes tam speciales quam generales, ratificavimus approbavimus et confirmavimus tenoreque presentis carte nostre ratificamus approbamus et confirmamus omnia et singula infeofamenta mortificationes donationes et dispositiones quascunque per nos dictamque nostram quondam charissimam matrem aliosque nostros preclarissimos progenitores, in favorem dictorum Prepositi Ballivorum et Consulum dicti Burgi nostri de Edinburgh, collegii scholarum et hospitalium ejusdem, confectas, pro ministerii curam apud ecclesias dicti Burgi inservientis, magistrorum regentium aliorumque professorum curam infra predictum Collegium et Scholas inservientium, pauperum scholasticorum, aliorumque pauperum seniorum decrepitarum indigentium personarum orphanorum et infantium parentibus orbatorum infra dictum burgum, sustentatione, de omnibus et singulis beneficiis terris tenementis annuis redditibus decimis fructibus redditibus emolumentis aliisque, particulariter et generaliter supra expressis, aut aliquibus parte seu partibus earundem, de quibuscunque data seu datis tenore seu contentis eadem existunt, in omnibus et singulis punctis passis capitibus articulis clausulis conditionibus et circumstantiis suis quibuscunque in eisdem contentis, secundum formas

lords of our Privy Council, our commissioners foresaid, now after our lawful and perfect age, and all our revocations as well special as general, have ratified, approved and confirmed, and by the tenor of our present charter ratify, approve and confirm, all and sundry infeftments, mortifications, gifts and dispositions whatsoever made by us and our said late dearest mother and others our most noble progenitors in favour of the said Provost, Bailies and Councillors of our said Burgh of Edinburgh, the College, Schools and Hospitals of the same, for the entertainment of the Ministry serving the cure at the churches of the said Burgh, of the masters, regents and other professors serving the cure within the foresaid College and Schools, of poor scholars and other poor, aged, decrepit, indigent persons, orphans and infants destitute of parents, within the said Burgh, of all and sundry benefices, lands, tenements, annual rents, teinds, fruits, rents and other emoluments particularly and generally above expressed, or any part or parts of the same, of whatsoever date or dates, tenor or contents, the same may be, in all and sundry their points, passes, heads, articles, clauses, conditions, and circumstances whatsoever therein contained, after the forms and tenors thereof

et tenores earundem, cum omnibus inde sequutis vel que desuper sequi possunt: Insuper nos pro bono fideli et gratuito servitio nobis nostrisque preclarissimis progenitoribus per Prepositum Ballivos Consules et Communitatem dicti Burgi nostri de Edinburgh eorumque predecessores, temporibus retroactis, prestito et impenso, proque assidua cura magno amore et affectione per nos ad dictum Burgum nostrum de Edinburgh habita, ejusdemque decoratione, proque honesta sustentatione ministrorum curas apud dictas ecclesias inservientium, et magistrorum regentium aliorumque professorum liberarum scientiarum infra Collegium et Scholas ejusdem Burgi, et pauperum hospitalium, aliarumque decrepitarum mutilatarum et indigentium personarum, orphanorum et infantum parentibus orbatorum, quotidie infra dictum Burgum crescentium, ac etiam pro diversis aliis bonis causis et considerationibus nos moventibus, cum avisamento et consensu predicto, DE NOVO dedimus concessimus, disposuimus, mortificavimus proque nobis et successoribus nostris pro perpetuo confirmavimus, tenoreque presentis carte nostre, de novo damus concedimus disponimus mortificamus proque nobis et successoribus nostris pro perpetuo confirmamus prefatis Preposito Ballivis, Consulibus et Communitati dicti Burgi nostri de Edinburgh, eorumque

with all that has followed or may follow thereupon. Moreover we for the good, true and thankful service done and rendered to us and to our most noble progenitors by the Provost, Bailies, Councillors and Community of our said Burgh of Edinburgh and their predecessors in times bygone, and for the earnest care and great love and affection borne by us to our said Burgh of Edinburgh, and for decoration of the same, and for the honest sustentation of the ministers serving the cures at the said churches, and of the masters, regents and others professors of liberal sciences within the College and Schools of the same Burgh, and of the poor of the hospitals and other decrepit, mutilated and indigent persons, orphans and infants destitute of parents, daily increasing within the said Burgh, and also for divers other good causes and considerations moving us, with advice and consent foresaid, we have of new given, granted, disponed, mortified, and for us and our successors for ever confirmed, and by the tenor of our present charter of new give, grant, dispone, mortify, and for us and our successors for ever confirm, to the foresaid Provost, Bailies, Councillors and Community of our said Burgh of Edinburgh and their successors in all time

successoribus omni tempore futuro, omnes et singulas predictas terras tenementa domos edificia hortos pomaria croftas ecclesias capellas annuos redditus fructus devorias proficua emolumenta census elimosinas lie dailsilver obitus et anniversaria quecunque, que quovismodo ad quascunque capellanias alteragia et prebendas in quacunque ecclesia capella aut collegio, infra libertatem dicti nostri Burgi, per quoscunque patronos eorundem fundatas, pertinuerunt seu pertinere dinoscuntur, in quorum possessione capellani et prebendarii earundem perprius fuerunt, ubicunque dicta tenementa domus edificia horti pomaria terre annui redditus anniversaria fructus proventus et emolumenta jacent aut per prius levata respective fuerunt, cum maneriorum locis pomariis terris annuis redditibus emolumentis et devoriis quibuscunque, que perprius ad Fratres Dominicales Predicatores, et Minores seu Franciscanos dicti Burgi nostri de Edinburgh pertinuerunt, unacum omnibus et singulis terris tenementis et domibus infra dictum Burgum et libertatem ejusdem jacentibus, cum omnibus annuis redditibus de quibuscunque domibus terris seu tenementis infra dictum Burgum levandis, quibuscunque capellaniis alteragiis ecclesiis mortuariis seu anniversariis fundatis datis et mortificatis, ubicunque eedem infra dictum

coming, all and sundry the foresaid lands, tenements, houses, buildings, yards, orchards, crofts, churches, chapels, annual rents, fruits, duties, profits, emoluments, maills, alms, dailsilver, obits and anniversaries whatsoever, which in anyway belonged or are known to belong to whatsoever chaplainries, altarages and prebends, founded in whatsoever church, chapel, or college within the liberty of our said Burgh, by whatsoever patrons of the same, in possession whereof the chaplains and prebendaries of the same formerly were, wherever the said tenements, houses, buildings, yards, orchards, lands, annual rents, anniversaries, fruits, provents and emoluments lie, or were formerly uplifted respectively, with manor places, orchards, lands, annual rents, emoluments and duties whatsoever, which formerly belonged to tho Friars Dominican Preachers and Minorites or Franciscans of our said Burgh of Edinburgh, together with all and sundry lands, tenements and houses lying within the said Burgh and liberty of the same, with all annual rents to be uplifted furth of whatsoever houses, lands, or tenements within the said Burgh, founded given and mortified to whatsoever chaplainries, altarages, churches, burials, or anniversaries, wherever tho

regnum nostrum Scotie jacent aut existunt, una etiam cum omnibus et singulis annuis redditibus aliisque devoriis usitatis et consuetis, aut que per quamcunque ecclesiam extra dictum Burgum, a prefatis Preposito et Ballivis ejusdem e communibus redditibus hujusmodi Burgi, pro suffragiis celebrandis, peti aut requiri possunt: Ac similiter totam et integram predictam preposituram Trinitatis Collegii prope dictum Burgum situate, integrasque prebendas eidem spectantes et incumbentes, una cum predictis ecclesiis parochialibus rectoriis et vicariis de Sowtray et Lempitlaw, ad dictam preposituram annexatis, cumque ecclesiis rectoriis et vicariis de Ormestoun, jacentis infra vicecomitatum nostrum de   , Kirkurde jacentis infra vicecomitatum nostrum de Peblis, et Weymis jacentis infra vicecomitatum nostrum de Fyiff, ex antiquo ad dictam ecclesiam de Sowtray annexatis, unacum omnibus et singulis aliis ecclesiis decimis garbalibus aliisque decimis fructibus redditibus glebis domibus edificiis hortis pomariis annuis redditibus, cum loco pomario et horto nuncupato Dingwall Castell, eidem spectantibus et incumbentibus, cum omnibus et singulis terris tenementis fructibus emolumentis communiis juribus casualitatibus proficuis devoriis tenentibus tenandriis et justis pertinentiis dicte prepositure quibuscunque,

same lie or exist within our said kingdom of Scotland; together also with all and sundry annual rents and other duties used and wont, or which may be asked or claimed by whatsoever church outwith the said Burgh from the foresaid Provost and Bailies of the same furth of the common rents of the same Burgh, for making of suffrages; And likewise all and whole the foresaid provostry of Trinity College situated near the said Burgh and the whole prebends pertaining and belonging to the same, together with the foresaid parish churches, parsonages and vicarages of Soltray and Lempitlaw annexed to the said provostry, and with the churches parsonages and vicarages of Ormiston lying within our sheriffdom of   , of Kirkurd lying within our sheriffdom of Peebles, and Wemyss lying within our sheriffdom of Fife, annexed of old to the said church of Soltray, together with all and sundry other churches, teind sheaves and other teinds, fruits, rents, glebes, houses, buildings, yards, orchards, annual rents, with the place, orchard and garden called Dingwall Castell pertaining and belonging to the same, with all and sundry lands, tenements, fruits, emoluments, commons, rights, casualties, profits, duties, tenants, tenandries

ubicunque eadem infra dictum regnum nostrum Scotie sunt aut jacent; Nec non totam et integram preposituram ecclesie Sancti Ægidii infra dictum nostrum Burgum situate, cum omnibus et singulis ecclesiis prebendis et capellaniis ad eandem spectantibus et pertinentibus, et specialiter ecclesias de Dumbarnie Pottie et Moncreif, rectorias et vicarias hujusmodi ad eandem ex antiquo annexatas, cum omnibus terris tenementis annuis redditibus domibus edificiis hortis decimis garbalibus aliisque decimis fructibus redditibus emolumentis juribus casualitatibus proficuis et devoriis tenentibus tenandriis et libere tenentium servitiis, ac justis suis pertinentiis quibuscunque, eisdem spectantibus et incumbentibus, ubicunque eadem sunt aut infra dictum regnum nostrum Scotie jacent; Et similiter totum et integrum locum monialem de Scheynis, vulgo nuncupatum the Nunrie of the Scheynis, jacentem infra libertatem et territorium dicti nostri Burgi, cum loco et hortis ejusdem, omnibusque terris tenementis domibus edificiis ecclesiis capellaniis prebendis decimis garbalibus aliisque decimis fructibus redditibus emolumentis casualitatibus juribus proficuis tenentibus tenandriis libere tenentium servitiis, et justis suis pertinentiis quibuscunque, ad eundem pertinentibus et spectantibus, ubicunque sunt aut infra dictum regnum

and just pertinents of the said provostry whatsoever, wherever the same are or lie within our said kingdom of Scotland. Also all and whole the provostry of the church of St Giles situated within our said Burgh, with all and sundry churches, prebends and chaplainries belonging and pertaining thereto, and specially the churches of Dumbarnie, Potie and Moncreif, parsonages and vicarages of the same annexed thereto of old, with all lands, tenements, annual rents, houses, buildings, yards, teind sheaves and other teinds, fruits, rents, emoluments, rights, casualties, profits and duties, tenants, tenandries and services of free tenants, and their just pertinents whatsoever, pertaining and belonging thereto, wherever the same are or lie within our said kingdom of Scotland; And likewise all and whole the nuns place of Scheynis commonly called the Nunrie of the Scheynis lying within the freedom and territory of our said Burgh, with the place and yards of the same, and all lands, tenements, houses, buildings, churches, chaplainries, prebends, teind sheaves and other teinds, fruits, rents, emoluments, casualties, rights, profits, tenants, tenandries and services of free tenants, and their just pertinents whatsoever,

nostrum Scotie jacent; Ac etiam totum et integrum hospitale operis Sancti Pauli, vulgo nuncupatum Sanct Paullis Work, jacens apud finem venelle, vulgo lie Leyth Wynd appellate, prope dictum Burgum nostrum de Edinburgh, ex parte orientali ejusdem venelle, cum omnibus terris tenementis annuis redditibus ecclesiis prebendis capellaniis decimis garbalibus aliis decimis fructibus redditibus emolumentis juribus casualitatibus proficuis tenentibus tenandriis libere tenentium servitiis et justis suis pertinentiis quibuscunque, eidem spectantibus et incumbentibus, ubicunque eadem infra dictum nostrum regnum Scotie sunt aut jacent, per prefatos Prepositum Ballivos Consules et Communitatem eorumque successores, pro sustentatione ministrorum curam apud ecclesias dicti Burgi inservientium, et seniorum decrepitorum orphanorum et pauperum infra eundem Burgum, hospitalium ejusdem, et pauperum scholasticorum infra Collegium et Scholas ejusdem, omni tempore affuturo, intromittendis levandis utendis et disponendis; Et similiter totam et integram predictam ecclesiam de Campis, nuncupatam lie Kirk of Field, cum acris loco et hortis eidem spectantibus et jacentibus, infra libertatem dicti nostri Burgi situatam, cum Archidiaconatu Lowthiane et ecclesia de Currie eidem annexatis, rectoria et vicaria

pertaining and belonging to the same, wherever the same are or lie within our said kingdom of Scotland; And also all and whole the hospital of the work of St Paul, commonly called Sanct Paullis Work, lying at the foot of the vennel commonly called Leyth Wynd near our said Burgh of Edinburgh, on the east side of the same vennel, with all lands, tenements, annual rents, churches, prebends, chaplainries, teind sheaves, other teinds, fruits, rents, emoluments, rights, casualties, profits, tenants, tenandries, services of free tenants, and their just pertinents whatsoever, pertaining and belonging to the same, wherever the same are or lie within our said kingdom of Scotland, to be intromitted with, uplifted, used, and disponed upon by the said Provost, Bailies, Councillors and Community and their successors for the sustentation of the ministers serving the cure at the churches of the said Burgh, and of the aged, decrepit, orphans and poor within the same Burgh of the hospitals of the same, and of the poor scholars within the College and Schools of the same, in all time coming; and likewise all and whole the foresaid Church of the Fields called the Kirk of Field with the acres, place and yards belonging and adjacent thereto, situated within the

ejusdem, unacum omnibus aliis ecclesiis capellaniis prebendis ad dictam ecclesiam de Campis et Archidiaconatum Lowthiane spectantibus, cum omnibus terris annuis redditibus tenementis domibus edificiis hortis decimis garbalibus aliis decimis fructibus redditibus proficuis proventibus emolumentis juribus casualitatibus et devoriis tenentibus tenandriis libere tenentium servitiis et justis pertinentiis quibuscunque, eisdem spectantibus et incumbentibus, ubicunque infra dictum regnum nostrum Scotie sunt aut jacent, per prefatos Prepositum Ballivos Consules et Communitatem eorumque successores, ad utilitatem et commodum predicti Collegii dicti nostri Burgi, magistrorum regentium aliorumque professorum curam infra Collegium prescriptum inservientium, omni tempore affuturo, intromittendis levandis utendis et disponendis, unacum omni jure titulo interesse et juris clameo, tam petitorio quam possessorio, que nos predecessores aut successores nostri habuimus habemus seu quovismodo habere vel clamare poterimus vel poterint in et ad terras annuos redditus ecclesias decimas aliaque specialiter et generaliter supra mentionata cum pertinentiis, aut aliquam earundem partem, sive ad census firmas proficua et devorias hujusmodi, de quibuscunque annis seu terminis preteritis seu futuris, ratione warde

liberty of our said Burgh, with the Archdeaconry of Lothian and church of Currie annexed thereto, parsonage and vicarage of the same, together with all other churches, chaplainries, [and] prebends belonging to the said church of the Fields and Archdeaconry of Lothian, with all lands, annual rents, tenements, houses, buildings, yards, teind sheaves, other teinds, fruits, rents, profits, provents, emoluments, rights, casualties and duties, tenants, tenandries, services of free tenants, and just pertinents whatsoever pertaining and belonging to the same, wheresoever they are or lie within our said kingdom of Scotland, to be intromitted with, uplifted, used and disponed upon by the foresaid Provost, Bailies, Councillors and Community, and their successors, for the utility and advantage of the foresaid College of our said Burgh, of the masters, regents and others professors serving the cure within the foresaid College in all time coming, together with all right, title, interest, and claim of right as well petitory as possessory, which we our predecessors or successors had, have, or in anyways could have or claim in and to the lands, annual rents, churches, teinds and others specially and generally above mentioned, with the pertinents, or any part thereof, or to the

relevii nonintroitus eschete forisfacture recognitionis reductionis disclamationis bastardie et tanquam ultimus heres aut ob non solutionem firmarum et devoriarum terrarum aliorumque prescriptorum, quorumcunque annorum seu terminorum preteritorum, diem date presentis carte nostre precedentium, aut defectus confirmationis debito in tempore, aut virtute acti annexationis omnium terrarum ecclesiasticarum ad coronam nostram, aut quorumcunque aliorum actorum legum statutorum aut dicti regni nostri Scotie constitutionum in contrarium factorum vel faciendorum, aut ob quamcunque aliam causam crimen aut occasionem preteritam, diem date presentis carte nostre precedentem, renunciando transferendo et extradonando eadem, cum omnibus actione et instantia earundem, pro nobis et successoribus nostris, in favorem prefatorum Prepositi Ballivorum Consulum et Communitatis eorumque successorum pronunc et in perpetuum, cum pacto de non petendo, ac cum supplemento omnium defectuum et imperfectionum, tam non nominatarum quam nominatarum, quas tanquam pro expressis in hac presenti carta nostra haberi volumus: Et similiter nos cum avisamento et consensu predicto univimus annexavimus et incorporavimus, tenoreque presentis carte nostre unimus annexamus et incorporamus

maills, fermes, profits and duties of the same, of whatsoever years or terms past or to come, by reason of ward, relief, nonentry, escheat, forfeiture, recognition, reduction, disclamation, bastardy, and as last heir, or for non-payment of maills and duties of the lands and others beforewritten of whatsoever years or terms byegone, preceding the date of our present charter, or of want of confirmation in due time, or by virtue of the act of annexation of all church lands to our crown, or of whatsoever other acts, laws, statutes, or constitutions of our said kingdom of Scotland made or to be made in the contrary, or for whatsoever other cause, crime, or occasion byegone, preceding the date of our present charter; renouncing, transferring, and overgiving the same, with all action and instance of the same, for us and our successors, in favour of the foresaid Provost, Bailies, Councillors and Community and their successors for now and for ever, *cum pacto de non petendo*, and with supplement of all defects and imperfections as well not named as named which we will to be held as expressed in this our present charter. And likewise we, with advice and consent foresaid, have united, annexed and incorporated, and by the tenor of our present charter unite, annex and

omnes et singulas predictas terras et tenementa domos edificia ecclesias preposituras prebendas capellanias hortos pomaria croftas annuos redditus decimas fructus proficua emolumenta aliaque particulariter et generaliter supra expressa cum earundem pertinentiis, in unum corpus, nunc et omni tempore futuro, nostram Fundationem Ministerii et Hospitalis Edinburgi nuncupandum; Ac volumus et concedimus ac pro nobis et successoribus nostris decernimus et ordinamus, quod unica sasina, nunc semel per presentes Prepositum et Ballivos dicti Burgi nostrido Edinburgh eorumve aliquem, apud Pretorium dicti nostri Burgi capienda, stabit et sufficiens erit sasina pro perpetuo omni tempore affuturo pro omnibus et singulis terris tenementis ecclesiis decimis aliisque specialiter et generaliter supra mentionatis, cum pertinentiis, absque aliqua alia speciali aut particulari sasina apud aliquam aliam partem seu partes hujusmodi, in futurum capienda, non obstante quod insimul et contigue minime jacent; quocirca nos pro nobis et successoribus nostris dispensavimus tenoreque presentis carte nostre dispensamus in perpetuum: TENENDAS ET HABENDAS omnes et singulas predictas terras tenementa domos edificia ecclesias capellas hortos pomaria croftas annuos redditus fructus devorias proficua emolumenta census eleemosinas lie daill silver obitus

incorporate all and sundry the foresaid lands and tenements, houses, buildings, churches, provostries, prebends, chaplainries, yards, orchards, crofts, annual rents, teinds, fruits, profits, emoluments and others particularly and generally above expressed, with the pertinents of the same, into one Body to be called now and in all time coming our FOUNDATION of the MINISTRY and HOSPITALITY of EDINBURGH; and we will and grant, and for us and our successors decern and ordain, that one sasine now once to be taken by the present Provost and Bailies of our said Burgh of Edinburgh, or any of them, at the Tolbooth of our said Burgh, shall stand and be sufficient sasine for ever in all time coming for all and sundry the lands, tenements, churches, teinds, and others specially and generally above mentioned, with the pertinents, without any other special or particular sasine to be taken in future at any other part or parts of the same, notwithstanding that they do not lie contiguous and together, whereanent we for us and our successors have dispensed and by the tenor of our present charter dispense for ever To HAVE AND TO HOLD all and sundry the foresaid lands, tenements, houses, buildings, churches, chapels, yards, orchards, crofts, annual rents, fruits, duties, profits, emoluments, maills, alms,

et anniversaria quecunque que quovismodo ad quascunque capellanias alteragia et prebendas in quacunque ecclesia capella aut collegio infra libertatem dicti nostri Burgi, per quoscunque patronos eorundem fundatas, pertinuerunt seu pertinere dinoscentur, in quorum possessione capellani et prebendarii earundem perprius fuerunt, ubicunque dicta tenementa domus edificia horti pomaria terre annui redditus anniversaria fructus proventus et emolumenta jacent aut perprius levata perprius fuerunt, cum maneriorum locis pomariis terris annuis redditibus emolumentis et devoriis quibuscunque, que perprius ad dictos Fratres Predicatores Dominicales, et Minores seu Franciscanos dicti nostri Burgi de Edinburgh pertinuerunt, unacum omnibus et singulis terris tenementis et domibus infra dictum Burgum et libertatem ejusdem jacentibus, cum omnibus annuis redditibus de quibuscunque domibus terris aut tenementis infra dictum Burgum levandis quibuscunque capellaniis alteragiis ecclesiis mortuariis seu anniversariis, fundatis datis et mortificatis, ubicunque eedem infra dictum regnum nostrum Scotie jacent aut existunt; una etiam cum omnibus et singulis annuis redditibus aliisque devoriis usitatis et consuetis, aut que per quamcunque ecclesiam extra dictum Burgum a prefatis Preposito et Ballivos ejusdem,

dailsilver, obits and anniversaries whatsoever, which in anyways pertained or are known to pertain to whatsoever chaplainries, altarages and prebends, in whatsoever church, chapel, or college within the liberty of our said Burgh, by whatsoever patrons of the same they were founded, in possession whereof the chaplains and prebendaries of the same formerly were, wheresoever the said tenements, houses, buildings, yards, orchards, lands, annual rents, anniversaries, fruits, provents, and emoluments lie or were formerly uplifted, with the manor places, orchards, lands, annual rents, emoluments and duties whatsoever, which formerly pertained to the said Friars Dominican Preachers and Minorites or Franciscans of our said Burgh of Edinburgh, together with all and sundry lands, tenements and houses lying within the said Burgh and liberty of the same, with all the annual rents to be uplifted furth of any houses, lands, or tenements within the said Burgh, founded, given and mortified to whatsoever chaplainries, altarages, churches, burials or anniversaries, wherever the same lie or are within our said kingdom of Scotland, together also with all and sundry annual rents and other duties used and wont, or which may

e communibus redditibus hujusmodi Burgi pro suffragiis celebrandis peti aut requiri possunt; nec non totam et integram predictam preposituram Trinitatis Collegii prope dictum Burgum situatam, integrasque prebendas eidem spectantes et incumbentes, una cum predictis ecclesiis parochialibus rectoriis et vicariis de Sowtray et Lempitlaw ad dictam preposituram annexatis, cumque ecclesiis parochialibus rectoriis vicariis de Ormestoun jacentis infra dictum vicecomitatum nostrum de , Kirkurde jacentis infra dictum vicecomitatum nostrum de Peblis, et Weymis jacentis infra dictum vicecomitatum nostrum de Fyff, ex antiquo ad dictam ecclesiam de Sowtray annexatis, unacum omnibus et singulis aliis ecclesiis decimis garbalibus aliisque decimis mansis glebis fructibus redditibus domibus edificiis hortis pomariis annuis redditibus cum loco pomario et horto nuncupato lie Dingwall Castell eidem spectante et incumbente, cum omnibus et singulis terris tenementis fructibus emolumentis juribus casualitatibus proficuis devoriis tenentibus tenandriis et justis pertinentiis dicte prepositure quibuscunque, ubicunque eadem infra dictum regnum nostrum Scotie sunt aut jacent: Ac etiam totam et integram predictam preposituram dicte ecclesie collegiate Sancti Egidii infra dictum nostrum Burgum situate,

be asked or claimed by whatsoever church outwith the said Burgh from the foresaid Provost and Bailies of the same, furth of the common rents of the said Burgh, for making of suffrages; As also all and whole the foresaid provostry of the Trinity College situated near the said Burgh, and whole prebends pertaining and belonging to the same, together with the foresaid parish churches, parsonages and vicarages of Soltray and Lempitlaw annexed to the said provostry, and with the parish churches, parsonages and vicarages of Ormiston lying within our said sheriffdom of , of Kirkurde lying within our said sheriffdom of Peebles, and Wemyss lying within our said sheriffdom of Fife, annexed of old to the said church of Soltray, together with all and sundry other churches, teind sheaves and other teinds, manses, glebes, fruits, rents, houses, buildings, yards, orchards, annual rents, with the place, orchard and yard called Dingwall Castell, pertaining and belonging to the same, with all and sundry lands, tenements, fruits, emoluments, rights, casualties, profits, duties, tenants, tenandries and just pertinents of the said provostry whatsoever, wherever the same are or lie within our said kingdom

cum omnibus et singulis ecclesiis prebendis et capellanis ad eandem pertinentibus et spectantibus, et presertim predictas ecclesias de Dumbarnie Potie et Moncreif, rectorias et vicarias earundem ad easdem ex antiquo annexatas, cum omnibus terris tenementis annuis redditibus domibus edificiis hortis decimis garbalibus aliis decimis fructibus redditibus emolumentis juribus casualitatibus proficuis et devoriis tenentibus tenandriis et libere tenentium servitiis ac justis suis pertinentiis quibuscunque ad easdem spectantibus et pertinentibus, ubicunque eedem infra dictum regnum nostrum Scotie sunt aut jacent; nec non totum et integrum predictum locum monialem de Scheynis, vulgo nuncupatum the Nunrie of the Scheynis, jacentem infra libertatem et territorium dicti nostri Burgi, cum loco et hortis ejusdem omnibusque terris tenementis domibus edificiis ecclesiis prebendis capellaniis decimis garbalibus aliis decimis fructibus redditibus emolumentis juribus casualitatibus proficuis tenentibus tenandriis libere tenentium servitiis et justis pertinentiis quibuscunque ad eundem pertinentibus et spectantibus, ubicunque sunt aut infra dictum regnum nostrum Scotie jacent; ac etiam totum et integrum predictum hospitale operis Sancti Pauli, vulgo nuncupatum Sanct Paullis Work, jacens apud finem dicte venelle

of Scotland; And also all and whole the foresaid provostry of the said collegiate church of St Giles situated within our said Burgh, with all and sundry churches, prebends, and chaplainries pertaining and belonging to the same, and specially the foresaid churches of Dumbarnie, Potie and Moncreif, parsonages and vicarages of the same, annexed thereto of old, with all lands, tenements, annual rents, houses, buildings, yards, teind sheaves, other teinds, fruits, rents, emoluments, rights, casualties, profits, and duties, tenants, tenandries and services of free tenants, and their just pertinents whatsoever, belonging and pertaining to the same, wheresoever the same are or lie within our said kingdom of Scotland; Also all and whole the foresaid nuns' place of Scheynis, commonly called the Nunrie of the Scheynis, lying within the freedom and territory of our said Burgh, with the place and yards of the same, and all lands, tenements, houses, buildings, churches, prebends, chaplainries, teind sheaves, other teinds, fruits, rents, emoluments, rights, casualties, profits, tenants, tenandries, services of free tenants, and just pertinents whatsoever pertaining and belonging to the same, wherever they are or lie within our said kingdom of Scotland; and

vulgo lie Leyth Wynd appellate, prope dictum Burgum nostrum de Edinburgh, ex parte orientali ejusdem venelle, cum omnibus terris tenementis annuis redditibus ecclesiis prebendis capellaniis decimis garbalibus aliis decimis fructibus redditibus emolumentis juribus casualitatibus proficuis tenentibus tenandriis libere tenentium servitiis et justis pertinentiis quibuscunque ad eandem spectantibus et pertinentibus, ubicunque infra dictum regnum nostrum Scotie eadem sunt aut jacent; per prefatos Prepositum Ballivos Consules et Communitatem eorumque successores, pro sustentatione dictorum ministrorum curam apud dictas ecclesias predicti nostri Burgi inservientium, et seniorum decrepitorum orphanorum et pauperum infra eundem Burgum, hospitalium ejusdem, pauperum scholasticorum infra Collegium et Scholas hujusmodi, omni tempore affuturo intromittendis levandis utendis et disponendis: Et similiter tenendas et habendas totam et integram predictam Ecclesiam de Campis, nuncupatam lie Kirk of Field, cum acris loco et hortis eidem spectantibus et jacentibus, infra libertatem dicti nostri Burgi situatis; unacum predicto Archidiaconatu Lowthiane et ecclesie de Currie eidem annexate, rectoria et vicaria ejusdem; unacum omnibus aliis ecclesiis capellaniis prebendis ad

also all and whole the foresaid hospital of the work of St Paul, commonly called Sanct Paullis Work, lying at the foot of the said vennel commonly called Leith Wynd, near our said Burgh of Edinburgh, on the east side of the said vennel, with all lands, tenements, annual rents, churches, prebends, chaplainries, teind sheaves, other teinds, fruits, rents, emoluments, rights, casualties, profits, tenants, tenandries, services of free tenants, and just pertinents whatsoever, pertaining and belonging to the same, wheresoever they are or lie within our said kingdom of Scotland, to be intromitted with, uplifted, used and disponed upon in all time coming, by the foresaid Provost, Bailies, Councillors and Community, and their successors, for the sustentation of the said ministers serving the cure at the said churches of our foresaid Burgh, and of the aged, decrepit, orphans and poor, within the same Burgh of hospitals of the same, [and] of poor scholars within the College and schools thereof. And likewise to have and to hold all and whole the foresaid Church of the Fields, called the Kirk of Field, with the acres, place, yards, belonging and adjacent thereto, situated within the liberty of our said Burgh; together with the foresaid Archdeaconry

dictam Ecclesiam de Campis et Archidiaconatum Lowthiane spectantibus, cum omnibus terris tenementis annuis redditibus domibus edificiis hortis decimis garbalibus aliisque decimis fructibus redditibus proficuis proventibus emolumentis juribus casualitatibus et devoriis tenentibus tenandriis libere tenentium servitiis et justis pertinentiis quibuscunque eisdem spectantibus et incumbentibus, ubicunque eadem infra dictum regnum nostrum Scotie sunt aut jacent, prefatis Preposito Ballivis Consulibus et Communitati eorumque successoribus, per ipsos, ad utilitatem et commodum predicti Collegii prefati Burgi, magistrorum regentium aliorumque professorum curam infra idem Collegium inservientium, omni tempore futuro intromittendis levandis utendis disponendis, de nobis et successoribus nostris in puram et perpetuam eleemosinam imperpetuum, per omnes rectas metas suas antiquas et divisas, prout jacent in longitudine et latitudine, in domibus edificiis boscis planis moris marcsiis viis semitis aquis stagnis rivulis pratis pascuis et pasturis molendinis multuris et eorum sequelis aucupationibus venationibus piscationibus petariis turbariis carbonibus carbonariis cuniculis cunicularîis columbis columbariis fabrilibus brasinis brueriis et genistis silvis nemoribus et virgultis lignis tignis lapicidiis lapide et calce cum curiis et

of Lothian and church of Currie annexed to the same, parsonage and vicarage thereof; together with all other churches, chaplainries [and] prebends belonging to the said Church of the Fields and to the Archdeaconry of Lothian, with all lands, tenements, annual rents, houses, buildings, yards, teind sheaves and other teinds, fruits, rents, profits, provents, emoluments, rights, casualties and duties, tenants, tenandries, services of free tenants, and just pertinents whatsoever pertaining and belonging to the same, wheresoever the same are or lie within our said kingdom of Scotland, to the foresaid Provost, Bailies, Councillors, and Community, and their successors, to be intromitted with, uplifted, used and disponed upon in all time coming by them, for the utility and profit of the foresaid College of the foresaid Burgh, masters, regents and other professors serving the cure within the same College, of us and our successors, in pure and perpetual alms for ever, by all their just ancient marches and divisions, as they lie in length and breadth, in houses, buildings, woods, plains, moors, marshes, ways, paths, waters, ponds, streams, meadows, pastures and pasturages, mills, multures and their sequels, hawkings, huntings,

earum exitibus herezeldis bluduitis et mulierum merchetis, cum libero introitu et exitu, ac cum omnibus aliis et singulis libertatibus commoditatibus proficuis et asiamentis ac justis suis pertinentiis quibuscunque, tam non nominatis quam nominatis, tam subtus terra quam supra terram, procul et prope ad predictam nostram fundationem aliaque premissa, cum pertinentiis spectantibus seu juste spectare valentibus quomodolibet in futurum, libere quiete plenarie integre honorifice bene et in pace sine aliqua revocatione contradictione aut obstaculo quocunque: Reddendo inde annuatim prefati Prepositus Ballivi Consules et Communitas dicti nostri Burgi eorumque successores, et pauperes dicti Hospitalis, ministri magistri et dicti Collegii regentes et scholarum, eorumque successores, devotas et humiles quotidianas preces Deo Omnipotenti pro preservatione nostri et successorum nostrorum, et sustentando ministros curam apud ecclesias respective predictas eorumque successores inservientes, aut solvendo pro eorum sustentatione tertiam partem fructuum et devoriarum ecclesiarum et beneficiorum superius mentionatorum, in optione Prepositi Ballivorum Consulum et Communitatis eorumque successorum tantum: In cujus rei testimonium huic presenti carte nostre magnum sigillum nostrum apponi precepimus.

fishings, peats, turfs, coals, coal pits, rabbits, warrens, doves, dove cots, forges, malt kilns, breweries, and heaths, woods, groves, thickets, wood, timber, stone quarries, stone and lime, with courts and their issues, herezelds, bludewites, marchets of women, with free ish and entry, and with all and sundry other liberties, commodities, profits, and easements, and their just pertinents whatsoever, as well not named as named, as well below the earth as above the earth, far or near, belonging or that could justly belong in any manner of way in future, to our foresaid foundation and others the premises, with the pertinents, freely, quietly, fully, wholly, honourably, well and in peace, without any revocation, contradiction, or obstacle whatsoever; Rendering therefor yearly, the foresaid Provost, Bailies, Councillors and Community of our said Burgh, and their successors, and the poor of the said hospital, the ministers, masters and regents of the said College and schools and their successors, devout and humble daily prayers to God Almighty for the preservation of us and our successors, and sustaining of the ministers serving the cure at the said churches respectively, and their successors, or paying for their sustentation the

TESTIBUS predilectis nostris consanguineis et consiliariis, Jacobo marchione de Hamiltoun comite Arranie domino Evan, Georgio comite Mariscalli domino Keyth, regni nostri Scotie mariscallo, Alexandro comite de Dunfermline domino Fyvie et Vrquhat &c. cancellario nostro, dilectis nostris familiaribus consiliariis, dominis Thoma Hamiltoun de Byris, nostro secretario, Ricardo Cokburne juniore de Clerkingtoun, nostri secreti sigilli custode, Alexandro Hay, nostrorum rotulorum registri ac consilii clerico, Joanne Cokburne de Ormestoun, nostre justiciarie clerico, militibus: magistro Joanne Scott de Caiplie, nostre cancellarie directore. Apud Beauvoir castell septimo die mensis Augusti anno Domini millesimo sexcentesimo duodecimo, regnorumque nostorum annis quadragesimo sexto et decimo, respective.

third part of the fruits and duties of the churches and benefices above mentioned, at the option of the said Provost, Bailies, Councillors and Community and their successors only. IN WITNESS whereof we have commanded our great seal to be affixed to this our present charter. WITNESSES, our well beloved cousins and councillors, James marquis of Hamilton earl of Arran lord Evan, George earl Marshal lord Keith marischal of our kingdom of Scotland, Alexander earl of Dunfermline lord Fyvie and Urquhart etc. our chancellor; our beloved familiar councillors, Sir Thomas Hamilton of Byres our secretary, Sir Richard Cockburn younger of Clerkington keeper of our privy seal, Sir Alexander Hay clerk of our rolls register and council, Sir John Cockburn of Ormiston our justice clerk, knights, Master John Scott of Caiplie director of our chancery. At Beauvoir Castle, the seventh day of the month of August, in the year of our Lord one thousand six hundred and twelve, and in the forty-sixth and tenth years of our reign respectively.

XVIII.

CHARTER granted by King Charles the First, under his Great Seal, to the Provost, Bailies, Councillors, and Community of the City of Edinburgh. Newmarket, 23d October 1636.

CAROLUS Dei gratia Magne Britannie Francie et Hibernie Rex, fideique defensor: Omnibus probis hominibus totius terre sue clericis et laicis salutem: SCIATIS nos memoria recolentes et perfecte intelligentes multa bona notabilia et gratuita servitia, per Prepositum Ballivos Consules Communitatem et Incolas Civitatis nostre Edinburgi, que est principalis et capitalis Civitas et Burgus antiqui regni nostri Scotie, prestita et impensa, non solum nobismetipsis a tempore felicis nostre successionis ad regnum, verum etiam quondam nostro charissimo patri eterne memorie et aliis nostris preclarissimis progenitoribus Scotie Regibus, quorum particulares et notabiles expressiones, in antiquis infeofamentis ipsis, per predecessores nostros eterne memorie concessis, contente sunt, que posteritati tanquam signa eorum fidelitatis, et ingentium et egregiorum servitiorum per ipsos bono et honori regni prestitorum et impensorum, remanent. Necnon nos considerantes, quod quedam questio

CHARLES by the grace of God, of Great Britain, France, and Ireland, King, and Defender of the Faith, To all good men of his whole land, clerics and laics, greeting: KNOW ye that we, calling to mind and perfectly understanding the many good, notable, and thankful services rendered and performed by the Provost, Bailies, Councillors, Community, and Inhabitants of our City of Edinburgh, which is the principal and capital City and Burgh of our ancient kingdom of Scotland, not only to us from the time of our happy succession to the kingdom, but also to our late dearest father of eternal memory, and to others our most noble progenitors, Kings of Scotland, of which particular and notable expressions are contained in ancient infeftments granted to them by our predecessors of eternal memory, which remain to posterity as memorials of their fidelity, and of the great and remarkable services rendered and performed by them to the advantage and honour of the realm. Also we, considering that a certain question has been raised in our said

facta dicte Civitati nostre Edinburgene penes amplitudinem et extentionem novi infeofamenti per dictum quondam nostrum charissimum patrem ipsis concessi, de data apud Halyruidhous [decimo] quinto die mensis Martii anno Domini millesimo sexcentesimo tertio, utrum is scrupulus et controversia que super dicta carta moveri poterit vel in eo quod tendere potest prejudicio nostro in particulari vel reipublice et regno in generali removeretur, ipsi voluntarie et ex ipsorum proprio motu omni cum humilitate in presentia dominorum nostri secreti consilii, vigesimo octavo die mensis Januarii anno Domini millesimo sexcentesimo trigesimo, comparuerunt et ibidem produxerunt in presentia dictorum dominorum, quoddam actum per Prepositum, Ballivos, Consules et Communitatem dicte nostre Civitatis factum, sub subscriptione ipsorum communis clerici, de data dicto vigesimo octavo die mensis Januarii anno Domini predicto, virtute cujus acti ratificarunt oblationem per dilectum nostrum consiliarium Dominum Joannem Hay nostri registri clericum, pro tempore designatum Magistrum Joannem Hay ipsorum clericum et commissionarium ipsorum nomine, penes restrictionem officii Vicecomitis et Coronatoris in dicto infeofamento concesso anno Domini millesimo sexcentesimo tertio predicto continentem

City of Edinburgh concerning the effect and extent of a new infeftment granted to them by our said late dearest father, of date at Holyrood house, the fifteenth day of the month of March, in the year of our Lord one thousand six hundred and three, and to the end that all scruple and controversy which can be raised concerning the said charter, which may tend either to our prejudice in particular or to that of the commonweal and kingdom in general, may be removed, they voluntarily and of their own free will appeared with all humility in presence of the lords of our Privy Council, on the twenty-eighth day of the month of January, in the year of our Lord one thousand six hundred and thirty, and there produced in presence of the said lords a certain Act made by the Provost, Bailies, Councillors, and Community of our said City, under the subscription of their Common Clerk, of date the said twenty-eighth day of the month of January, in the year of our Lord foresaid, by virtue of which act they ratified the offer made by our beloved councillor, Sir John Hay, clerk of our register, for the time designed Master John Hay, their clerk and commissioner in their name, concerning the restriction of the office of Sheriff and Coroner contained in the said infeftment, granted in the year of our Lord

et jurisdictionem ejusdem cum tentione suarum gilde curiarum, ad boudas subsequentes continentem in oblatione per dictam nostram Civitatem nobilibus et generosis Lothiane occidentalis facta, que comprehendit terras subsequentes, viz. Burgum nostrum de Edinburgh, communia molendina, communem moram, commune marresium lie myre, lacus, menia et fossas ejusdem, villas de Leith et Newhevin, portus, navium, stationes, propugnacula lie peires schoires raides lie linkis eister et wester ejusdem, terras nuncupatas lie Commoun Clossetts Burssis Hoilf Hallis et alias terras ad dictum Burgum nostrum proprie spectantes vias, semitas, stratas, plateas et passagia ad dictum nostrum Burgum, et ab eodem, et ad Leith et Newhevin, vel ab eisdem respective, ducentes, per eorum communem moram et marresium, et que per ipsos constructa seu reparata sunt, quousque dicta mora et marresium extendunt, cum viis et plateis ducentibus ad ipsorum communia molendina super aquam de Leith, et ab iisdem quousque dicta molendina, cum terris molendinariis ad dictum nostrum Burgum proprie spectantibus ibidem extendunt; Ac etiam penes renunciationem in favorem nostrum omnis juris regalitatis ipsis vel eorum predecessoribus dispositi, si quod sit, cum omni jure quod habent ad eschaetas personarum criminalium

one thousand six hundred and three foresaid, and jurisdiction of the same, with the holding of their Guild courts, to the bounds after set forth, contained in an offer by our said City to the nobles and gentry of West Lothian, which comprehends the lands following, namely, our Burgh of Edinburgh, common mills, common muir, common march, or myre, loch, walls and fosses of the same, towns of Leith and Newhaven, ports, stations of ships, bulwarks, peires, schoires, raides, links eister and wester of the same, the lands called the commoun clossetts, burssis, hoilf hallis, and other lands properly belonging to our said Burgh, ways, paths, streets, places and passages leading to our said Burgh and from the same, and to Leith and Newhaven, or from the same respectively, through their common muir and marsh, and which have been either constructed or repaired by them so far as the said muir and marsh extend, with the ways and places leading to their common mills upon the Water of Leith, and from the same as far as the said mills with mill lands properly belonging to our said Burgh there extend. And also concerning the renunciation in our favour of all right of regality granted to them or their predecessors, if such there be, with all right which

coram ipsis morti convictarum; necnon eschaetarum quarumcunque incolarum dicti Burgi nostri ad cornu pro civilibus causis denunciatarum, cum omnibus aliis extraordinariis libertatibus, si que sint, in dicto infeofamento contentis, que subdito non competent nec conceduntur, nec hucusque concesse fuerunt vicecomitatibus coronatoribus seu liberis Burgis; necnon penes renunciationem omnis juris quod ad australem seu borealem ripas lie bankis Castri nostri de Edinburgh habent, absque prejudicio omnimodo alicujus prioris legitimi juris quod dictus Burgus aut aliqui burgenses ejusdem ad easdem habent, et absque innovatione usus dicte australis ripe conformiter prout omnibus temporibus retroactis, ultra hominum memoriam consuetum fuit; necnon continente humilem petitionem quatenus nobis placeret ratificare dictam cartam et ipsorum universa antiqua infeofamenta inibi contenta, cum libera libertate et sola seu unica negociatione lie sole tread infra bondas Lothiane occidentalis ad liberum burgum regale pertinente, unacum dictis officiis Vicecomitis et Coronatoris infra bondas dicti Burgi nostri de Edinburgh, dictas villas de Leith et Newhevin, aliasque supra et subtus specificatas, unacum custumis minutis custumis et aliis inibi contentis, quorum dictus

they have to escheats of criminals condemned before them to death; also of the escheats of whatsoever inhabitants of our said Burgh are put to the horn for civil causes, with all other extraordinary liberties, if such there be, contained in the said infeftment, which are not competent nor granted to a subject, nor have hitherto been granted to sheriffs, coroners, or free burghs; also concerning the renunciation of all right which they have to the south or north banks of our Castle of Edinburgh, without prejudice in any way to any prior lawful right which the said Burgh or any burgesses of the same have to the said banks, and without any change in the use of the said south bank in conformity with the practice in all time by past, beyond the memory of man; also containing a humble petition that we would be pleased to ratify the said charter, and all their ancient infeftments therein contained, with free liberty, and the sole or exclusive trade "sole tread" within the bounds of West Lothian pertaining to a free royal burgh, together with the said offices of Sheriff and Coroner within the bounds of our said Burgh of Edinburgh, the said towns of Leith and Newhaven, and others above and under specified, together with the customs, petty customs, and others therein contained, of which the said Burgh was in posses-

Burgus fuit in possessione, unacum ceteris libertatibus privilegiis et aliis in dicto infeofamento contentis, tenore ejusdem conformiter; Quemadmodum, dicto acto conformiter, Prepositus, Ballivi, Consules et Communitas dicti Burgi nostri plenariam et amplam renunciationem et resignationem premissorum, in favorem nostrum nostrorumque successorum, fecerunt et subscripserunt, et nos ex nostris gratioso respectu et favore erga dictam nostram Civitatem Edinburgenam, et ad ipsos incitandos in ipsorum bono et gratuito servitio nobis et successoribus nostris continuare, acceptare volentes dictam ipsorum renunciationem, ac ratificare dicta sua infeofamenta, cum universis libertatibus et privilegiis inibi contentis, exceptis in particularibus supra mentionatis ut dictum est renunciatis, igitur nos post nostram legitimam et perfectam etatem viginti quinque annorum completam, et post omnes nostras revocationes, cum avisamento et consensu predilecti nostri consanguinei et consiliarii Joannis comitis de Traquair domini Lintoun et Caverstoun, &c. nostri supremi thesaurarii, computorum nostrorum rotulatoris, collectoris novarumque nostrarum augmentationum regni nostri Scotie, necnon cum avisamento et consensu reliquorum dominorum nostri Scaccarii dicti regni nostri Scotie

sion, together with the other liberties, privileges, and others contained in the said infeftment, conform to the tenor thereof. And whereas the Provost, Bailies, Councillors and Community of our said Burgh have, conform to the said act, made and subscribed a full and ample renunciation and resignation of the premises in favour of us and of our successors, and we, out of our gracious respect and favour towards our said City of Edinburgh, and as an incitement to them to continue in their good and thankful service to us and our successors, have been pleased to accept their said renunciation, and to ratify their said infeftments with all liberties and privileges therein contained, except in the particulars above mentioned, renounced as aforesaid, therefore after our lawful and perfect age of twenty-five years complete, and after all our revocations, with the advice and consent of our beloved cousin and councillor, John earl of Traquair, lord Lintoun and Caverstoun, etc., our High Treasurer, our comptroller and collector of our new augmentations of our kingdom of Scotland, as also with advice and consent of the rest of the lords of our Exchequer of our said kingdom of Scotland, our commissioners, have ratified, approved, and by this our present charter have confirmed, and by the tenor

nostrorum commissionariorum ratificavimus approbavimus et hac presenti carta nostra confirmavimus, tenoreque ejusdem ratificamus approbamus ac pro nobis et successoribus nostris pro perpetuo confirmamus in favorem Prepositi, Ballivorum, Consulum et Communitatis dicte Civitatis nostre Edinburgene, eorumque successorum, particulares et severales cartas, infeofamenta, donationes et alia subscripta, viz. CARTAM factam et concessam, per quondam Regem Robertum Bruce, Scotie Regem.

. . . .

Et similiter CARTAM factam et concessam per dictam quondam nostram charissimam aviam Reginam Mariam, sub suo magno sigillo, dictis Preposito, Ballivis, Consulibus et Communitati dicti nostri Burgi, et eorum successoribus, de omnibus et singulis terris, tenementis, domibus, edificiis, ecclesiis, capellaniis, hortis, pomariis, croftis, annuis redditibus, decimis, servitiis, proficuis, devoriis, emolumentis firmis lie almous daillsilver obites et anniversariis, que ad quascunque capellanias, alteragia et prebendarias in quibuscunque ecclesiis, capellis seu collegiis, infra libertatem dicti nostri Burgi, per quemcunque patronum fundatas, pertinuerunt, in quorum possessione dicti capellanii seu prebendarii fuerunt, ubicunque eadem jacent, infra dictum regnum nostrum Scotie, vel extra vel intra dictum nostrum Burgum de Edinburgh, et omnium

of the same ratify, approve, and for us and our successors for ever confirm, in favour of the Provost, Bailies, Councillors and Community of our said City of Edinburgh and their successors, the particular and several charters, infeftments, gifts, and others underwritten, videlicet, A CHARTER made and granted by the former King Robert Bruce, King of Scotland.

. . . .

And likewise a CHARTER made and granted by our said late dearest grandmother Queen Mary under her great seal, to the said Provost, Bailies, Councillors and Community of our said Burgh, and their successors, of all and sundry lands, tenements, houses, buildings, churches, chaplainries, yards, orchards, crofts, annual rents, teinds, services, profits, duties, emoluments, rents, almous, daillsilver, obites, and anniversaries, which belonged to whatsoever chaplainries, altarages, and prebends, founded in whatsoever churches, chapels, or colleges within the liberty of our said Burgh, by whatsoever patron, in possession whereof the said chaplains or prebendaries were, wheresoever the same lie within our said kingdom of Scotland, either without or within our said Burgh of Edinburgh, and of all lands which

terrarum que ad fratres Dominicanos et Franciscanos vulgo the Blackfrieris et Grayfrieris pertinuerunt, cum diversis aliis in dicta carta plenius specificatis, que est de data, decimo tertio die mensis Martii, anno Domini millesimo quingentesimo sexagesimo sexto. Ac etiam CARTAM concessam per dictum quondam nostrum charissimum patrem, sub suo magno sigillo, de data apud Striveling, decimo quarto die mensis Aprilis, anno Domini millesimo quingentesimo octogesimo secundo, per quam ratificat et confirmat dictam priorem cartam, concessam per dictam quondam suam charissimam matrem, de dictis terris et aliis predictis, de data dicto decimo tertio die mensis Martii, anno Domini millesimo quingentesimo sexagesimo sexto, necnon ratificat et approbat dimissionem et renunciationem factam per Joannem Gib in favorem dicti Burgi nostri de Edinburgh, de ipsius jure ad prepositurам de Kirkafield, universas domos, terras et edificia ad eandem pertinentes; et per quam dictus quondam noster charissimus pater dedit, concessit et disposuit dictis Preposito, Ballivis, Consulibus et Communitati dicti nostri Burgi de Edinburgh, eorumque successoribus, licentiam et libertatem erigendi Collegium, construendi et reparandi sufficientes domos et loca, pro receptione professorum humanitatis, literarum et

belonged to the Friars Dominican and Franciscan, commonly [called] the Blackfriars and Greyfriars, with sundry others more fully set forth in the said charter, which is dated the thirteenth day of the month of March, in the year of our Lord one thousand five hundred and sixty-six. As also a CHARTER granted by our said late dearest father, under his great seal, dated at Stirling, the fourteenth day of the month of April, in the year of our Lord one thousand five hundred and eighty-two, by which he ratifies and confirms the said former charter, granted by his said late dearest mother of the said lands and others foresaid, of date the said thirteenth day of the month of March, in the year of our Lord one thousand five hundred and sixty-six, also he ratifies and approves the demission and renunciation made by John Gib in favour of our said Burgh of Edinburgh of his right to the provostry of Kirk of Field, whole houses, lands, and buildings pertaining to the same; and by which our said late dearest father gave, granted, and disponed to the said Provost, Bailies, Councillors and Community of our said Burgh of Edinburgh and their successors permission and liberty to erect a College, to construct and repair suitable houses and places for the reception of professors

linguarum philosophie, theologie, medicine, legum, omniumque aliarum scientiarum liberalium, ac eligendi sufficientes professores ad dictas professiones docendas, et ad hunc effectum, dedit concessit et disposuit ipsis et eorum successoribus, preposituram de lie Kirk of Field cum terris, tenementis, fructibus, possessionibus, redditibus, devoriis et pertinentiis ejusdem. Necnon CARTAM factam et concessam per dictum quondam nostrum charissimum patrem, sub suo magno sigillo, de data quarto die mensis Aprilis, anno Domini millesimo quingentesimo octogesimo quarto, per quam dictus quondam noster charissimus pater, considerans quod dicti Prepositus, Ballivi, Consules et Communitas dicti nostri Burgi de Edinburgh magnos sumptus et expensas impenderunt in erectione dicti Collegii, domorum constructione, et magnas monete summas dotarunt pro sustentatione professorum humanitatis, philosophie et theologie, infra idem, pro instructione juventutis inibi, dedit, concessit et disposuit dictis Preposito, Ballivis, Consulibus et Communitati dicti nostri Burgi, ad usum dicti Collegii, ac pro sustentatione rectoris et regentium infra idem, totum et integrum Archidiaconatum Lothiane, inibi continentem rectoriam de Currie, cum manso, gleba,

of humanity, letters and languages, philosophy, theology, medicine, laws, and all other liberal sciences, and to elect competent professors to teach the said professions; and to that effect gave, granted, and disponed to them and their successors the provostry of the Kirk of Field with the lands, tenements, fruits, possessions, rents, duties, and pertinents of the same; also a CHARTER made and granted by our said late dearest father under his great seal, of date the fourth day of the month of April in the year of our Lord one thousand five hundred and eighty-four, by which our said late dearest father, considering that the said Provost, Bailies, Councillors and Community of our said Burgh of Edinburgh incurred great charges and expenses in the erection of the said College, [and] in the building of the houses, and gave great sums of money for the sustentation of professors of humanity, philosophy and theology within the same for the instruction of youth therein, gave, granted and disponed to the said Provost, Bailies, Councillors, and Community of our said Burgh, for the use of the said College, and for the support of the rector and regents within the same, all and whole the archdeaconry of Lothian, containing therein the parsonage of Curry, with the manse, glebe, church lands, teinds, fruits, rents, profits

terris ecclesiasticis, decimis, fructibus, redditibus, proficuis et devoriis ejusdem. Et similiter CARTAM factam et concessam per dictum quondam nostrum charissimum patrem, sub suo magno sigillo, de data, vigesimo sexto die mensis Maii, anno Domini millesimo quingentesimo octogesimo septimo, per quam dictus quondam noster charissimus pater considerans magnos sumptus et expensas per dictos Prepositum, Ballivos, Consules et Communitatem dicti nostri Burgi de Edinburgh impensas, in erectione hospitalis, in sustentatione suorum ministrorum et Collegii predictorum dedit, concessit et disposuit dictis Preposito, Ballivis, Consulibus et Communitati dicti Burgi nostri de Edinburgh, eorumque successoribus, totam et integram preposituram Collegii Trinitatis, terras, domos, redditus, ecclesias, aliosque fructus, redditus et emolumenta eidem annexata. Ac etiam CARTAM factam et concessam per dictum quondam nostrum charissimum patrem, sub suo magno sigillo, de data, vigesimo nono die mensis Julii, anno Domini millesimo quingentesimo octogesimo septimo, ratificantem infeofamenta concessa per seipsum ac per dictam quondam Mariam Reginam, ipsius matrem, de dictis terris ecclesiasticis Collegii Trinitatis, prepositure de Kirkafield, et Archidiaconatus de Louthiane, necnon continentem novam donationem, de omnibus

and duties of the same; and likewise a CHARTER made and granted by our said late dearest father, under his great seal, of date the twenty-sixth day of the month of May in the year of our Lord one thousand five hundred and eighty-seven, by which our said late dearest father, considering the great charges and expenses incurred by the said Provost, Bailies, Councillors and Community of our said Burgh of Edinburgh, in the erection of an Hospital, in sustaining their ministers and College foresaid, gave, granted and disponed to the said Provost, Bailies, Councillors and Community of our said Burgh of Edinburgh and their successors, all and whole the provostry of Trinity College, the lands, houses, rents, churches, and other fruits, rents and emoluments annexed to the same; and also a CHARTER made and granted by our said late dearest father, under his great seal, of date the twenty-ninth day of the month of July in the year of our Lord one thousand five hundred and eighty-seven, ratifying the infeftments granted by himself and by the said late Queen Mary, his mother, of the said church-lands of Trinity College, provostry of Kirk of Field, and archdeaconry of Lothian, also containing a new gift of all the said church lands, provostry, Trinity

dictis terris ecclesiasticis, prepositura, Collegio Trinitatis, et Archidiaconatu Lothiane, et de universis terris et decimis ad easdem pertinentibus, ad usum ministrorum, Collegii et pauperum. Ac etiam CARTAM factam et concessam per dictum quondam nostrum charissimum patrem, sub suo magno sigillo, de data, apud Beauvoir Castell, septimo die mensis Augusti, anno Domini millesimo sexcentesimo duodecimo continentem ratificationem omnium priorum jurium, dicto Burgo nostro de Edinburgh confectorum, de dictis terris ecclesiasticis, prepositura de Kirkafield, prepositura Collegii Trinitatis et Archidiaconatu Louthiane, unacum nova donatione dictarum integrarum terrarum ecclesiasticarum et beneficiorum, aliorumque inibi contentorum, applicandorum per dictos Prepositum et Ballivos eorumque successores, ad sustentationem suorum ministrorum collegii et pauperum respective: In omnibus et singulis punctis, passis, capitibus, articulis, clausulis, conditionibus, privilegiis, libertatibus, immunitatibus et circumstantiis quibuscunque in dictis infeofamentis, cartis, donationibus et aliis particulariter supraspecificatis et eorum unoquoque contentis secundum formas et tenores earundem. Proviso quod ratificatio dictarum cartarum, infeofamentorum et aliorum predictorum nullatenus comprehendet nec extendetur ad quod-

College, and archdeaconry of Lothian, and of the whole lands and teinds belonging to the same, for the use of the ministers, college and poor; and also a CHARTER made and granted by our said late dearest father, under his great seal, of date at Beauvoir Castle, the seventh day of the month of August, in the year of our Lord one thousand six hundred and twelve, containing a ratification of all prior rights granted to our said Burgh of Edinburgh, of the said church lands, provostry of Kirk of Field, provostry of Trinity College, and archdeaconry of Lothian, together with a new gift of the said whole church lands and benefices and others therein contained, to be applied by the said Provost and Bailies and their successors to the sustentation of their ministers, College, and poor respectively: IN ALL and sundry points, passes, heads, articles, clauses, conditions, privileges, liberties, immunities, and circumstances whatsoever contained in the said infeftments, charters, gifts, and others particularly above specified, and in every one of them according to the forms and tenors of the same. Providing that the ratification of the said charters, infeftments and others foresaid shall nowise comprehend nor be extended to any right of regality, if any such be com-

cunque jus regalitatis, si quod sit in dictis infeofamentis et juribus confirmatis, eorumve aliquo comprehensum, nec ultra extendetur penes hereditarium officium Vicecomitis et Coronatoris, et jurisdictionem hujusmodi, nec ad tentionem curiarum gilde, sed solummodo, infra boudas dicti Burgi nostri de Edinburgh, communia molendina, communem moram, commune marresium, lacus, menia et fossas ejusdem, et intra villas de Leith et Newhevin, portus, receptaculum propugnacula lie schoires peires raides lie linkis eister et wester, terras nuncupatas, lie commoun clossettis, Burse Hoilff Hallis et alias terras ad dictum Burgum nostrum in proprietate spectantes, vias, semitas, stratas plateas et passagia ad et a dicto Burgo nostro de Edinburgh ad et a dictis villis de Leith et Newhevin, et plateas vias et passagia ducentes per ipsorum communem moram et marresium, et que per ipsos construuntur et reparantur, quousque dicte mora et marresium extendunt, et ipsorum vias et plateas ducentes ad et ab ipsorum communis molendinis, super dictam aquam de Leith, quousque dicta molendina, terre molendinarie et omnes ad ipsos proprie spectantes extendunt. Et quod ratificatio dictarum cartarum nullatenus inferret impedimentum nec prejudicium nobis et successoribus nostris in nostro jure et prerogativa regali,

prehended in the said infeftments and confirmed rights, or in any of them, nor farther shall be extended to the heritable offices of Sheriff and Coroner and jurisdiction of the same, nor to the holding of courts of Guild, but only within the bounds of our said Burgh of Edinburgh, common mills, common moor, common marsh, lochs, walls and ditches of the same, and within the towns of Leith and Newhaven, ports, harbour, bulwarks, schoires, peires, raidis, linkis easter and wester, lands called the commoun clossettis, Burse Hoilff Hallis, and other lands belonging to our said Burgh in property, ways, paths, streets, places, and passages to and from our said Burgh of Edinburgh, and to and from the said towns of Leith and Newhaven, and places, ways, and passages leading through their common moor and marsh, and which were constructed and repaired by them, as far as the said moor and marsh extend, and their ways and places leading to and from their common mills upon the said water of Leith, as far as the said mills, mill lands, and all belonging to them properly extend. And that the ratification of the said charters shall nowise infer impediment nor prejudice to us and our successors in our right and royal prerogative to erect

ad erigendum burgos regalitatis seu burgos baronie in aliqua alia parte Lothiane occidentalis, extra bondas dicti Vicecomitatus et Coronatoriatus, ad quas idem modo predicto restringitur, cum hac declaratione omnimodo, quod restrictio dictorum officiorum Vicecomitis et Coronatoris et aliorum predictorum ad bondas supraspecificatas, nullum inferret prejudicium dicto Burgo nostro de Edinburgh et incolis ejusdem penes privilegium et libertatem sole seu unice negotiationis mercature infra universas bondas Lonthiane occidentalis, ad ipsos tanquam liberum burgum regalitatis pertinentis. Et similiter quod dicta ratificatio infeofamentorum cartarumque supraspecificatorum nullatenus extendetur ad quodcunque jus, quod dicta urbs et eorum successores inde habent, seu pretendere poterint ad eschaetas criminalium personarum, coram ipsis morti convictarum, neque ad eschaetas quarumcunque incolarum, infra bondas dictorum suorum Vicecomitatus et Coronatoriatus, ut dictum est restrictorum hactenus denunciatis, aut qui imposterum denunciati erunt ad cornu vel ob criminales vel civiles causas; Necnon quod dicta ratificatio nullatenus extendetur ad australem et borealem ripas lie bankis Castri nostri de Edinburgh, absque prejudicio omni-

burghs of regality or burghs of barony in any other part of West Lothian beyond the bounds of the said Sheriffship and Coronership, to which the same is restricted in manner foresaid, with this declaration always, that the restriction of the said offices of Sheriff and Coroner and others foresaid, to the bounds above specified, shall infer no prejudice to our said Burgh of Edinburgh and inhabitants of the same, concerning their privilege and liberty of the sole or exclusive trade of merchandise within the whole bounds of West Lothian belonging to them as a free burgh of regality And likewise that the said ratification of the infeftments and charters above specified shall in noways be extended to any right that the said City and their successors thence have or can pretend to the escheats of criminals condemned before them to death, nor to the escheats of any of the inhabitants within the bounds of their said Sheriffship and Coronership, restricted as said is, already put or who shall afterwards be put to the horn either for civil or criminal causes; also that the said ratification shall nowise be extended to the south and north banks of our Castle of Edinburgh ,without prejudice always to any lawful right that

modo cujuscunque juris legitimi, quod ipsi seu burgenses dicti nostri burgi ad easdem habent quintum diem mensis Martii, anno Domini millesimo sexcentesimo tertio, precedentem, et absque innovatione aut prejudicio dicto nostro Burgo et eorum successoribus, de usu dicte australis ripe lie bank conformiter prout solitum fuit, omnibus temporibus retroactis, ultra hominum memoriam; quemadmodum declaratur, quod particularia supraspecificata a dicta ratificatione excepta sunt, et tenebuntur tanquam excepta, non solum e dictis cartis et infeofamentis per presentis carte nostre tenorem ratificatis, verum etiam e quibuscunque aliis infeofamentis, juribus aut titulis factis et concessis, per predecessores nostros dicto nostro Burgo, pro quocunque tempore diem date presentis carte nostre precedente; absque prejudicio predilecto nostro consanguineo et consiliario Jacobo Lennocie Duce, etc. supremo regni nostri Scotie Admirallo et Camerario, heredibus suis et successoribus, in jure dictorum officiorum, de quocunque jure, titulo seu clameo que habere poterint ad quascunque libertatum privilegiorum jurisdictionum commoditatum aut aliorum quorumcunque contentorum in infeofamentis supraspecificatis concessis dicto nostro Burgo de Edinburgh, et eorum predecessoribus, et per nos in ipsorum favorem ratificatorum, prout de

they or the burgesses of our said Burgh have to the same, previous to the fifth day of the month of March, in the year of our Lord one thousand six hundred and three; and without innovation or prejudice to our said Burgh and their successors of the use of the said southern bank, conform to the practice of time past beyond the memory of man. Moreover, it is declared that the particulars above specified are excepted from the said ratification, and shall be held as excepted not only from the said charters and infeftments ratified by the tenor of this our present charter, but also from whatsoever other infeftments, rights, or titles made and granted by our predecessors to our said Burgh, at any time preceding the day of the date of our present charter; without prejudice to our beloved cousin and councillor James duke of Lennox, etc., High Admiral and Chamberlain of our kingdom of Scotland, and his heirs and successors in right of the said offices, of whatsoever right, title, or claim which they can have to any of the liberties, privileges, jurisdictions, commodities or others whatsoever, contained in the infeftments above specified, granted to our said Burgh of Edinburgh and their predecessors, and ratified by us in their favour as accords of law. And that the said

jure congruit. Et quod dictus Admirallus et Camerarius suique predicti et dicta Civitas nostra Edinburgensis et eorum successores erint in iisdem statu et conditione, penes omnia jura et privilegia dicta officia tangentia sicuti ante diem date presentis carte nostre erant, et sicuti presens nostra ratificatio nunquam concessa fuisset. Et similiter nos cum avisamento et consensu predicto, volumus et concedimus, ac pro nobis et successoribus nostris decernimus et ordinamus, quod hec presens nostra ratificatio cartarum, infeofamentorum, donationum, aliorumque particulariter supraspecificatorum, cum restrictionibus omnimodo exceptionibus, reservationibus et provisionibus suprascriptis, que in hac presenti carta nostra pro repetitis habentur, sit et erit tanti valoris, roboris, efficacie et effectus, in omnibus respectibus, ac si predicta universa infeofamenta, carte, donationes et alia prescripta ad longum et verbatim in hac presenti carta nostra insererentur, non obstante, eorundem minime insertione, penes quorum non insertionem et omnia desuper sequuta vel quo desuper sequi poterint, nos pro nobis et successoribus nostris dispensavimus, ac per presentis carte nostre tenorem, dispensamus in perpetuum. Necnon nos cum avisamento et consensu predicto, tenore presentis carte nostre, ratificamus et appro-

Admiral and Chamberlain and his foresaids, and our said City of Edinburgh and their successors shall be in the same state and condition, concerning all the rights and privileges touching the said offices, as they were in before the day of the date of our present charter, and as if our present ratification had never been granted. And in like manner we, with advice and consent foresaid, will and grant, and for us and our successors decern and ordain that this our present ratification of the charters, infeftments, gifts, and others particularly above specified, always with the restrictions, exceptions, reservations, and provisions above written, which are held as repeated in this our present charter, is and shall be of as much force, strength, efficacy and effect in all respects as if the whole foresaid infeftments, charters, gifts, and others foresaid were inserted at length and word for word in this our present charter, notwithstanding the not insertion of the same, concerning the not insertion of which and all following or that can follow thereupon, we for us and our successors have dispensed, and by the tenor of our present charter, dispense for ever. Likewise we, with advice and consent foresaid, by the tenor of our present charter, ratify and approve the heads and

bamus capita et articulos concordationis confecta inter dictos Prepositum, Ballivos, Consules et Communitatem dicte Civitatis nostre Edinburgene et ministros intra dictam Civitatem, in acto nostri secreti concilii eatenus confecto de data, primo die mensis Novembris, anno Domini millesimo sexcentesimo vigesimo quinto, contenta, et dictum actum concilii, de data predicta, eatenus confectum, in omnibus et singulis punctis clausulis et conditionibus inibi contentis; et specialiter absque prejudicio predicte generalitatis, illam clausulam ejusdem per quam concordatum fuit quod dicti Prepositus, Ballivi, Consules et Communitas dicte Civitatis nostre Edinburgene, eorumque successores, habebunt jus titulum et privilegium presentationis et nominationis ministrorum ad inserviendum curas universarum ecclesiarum infra dictum Burgum nostrum de Edinburgh, omnibus temporibus affuturis; ac volumus, concedimus, declaramus et ordinamus, quod presens hec nostra ratificatio dictorum capitum et articulorum concordationis particulariter et generaliter supraspecificatorum sit et erit tanti valoris, roboris, efficacie et effectus in omnibus respectibus, sicuti eadem ad longum in hac presenti carta nostra insererentur, penes quorum non insertionem, nos pro nobis et successoribus nostris, tenore presentis carte nostre dispensamus in per-

articles of agreement made between the said Provost, Bailies, Councillors and Community of our said City of Edinburgh, and the ministers within the said City, in an act of our privy council made concerning the same, of date the first day of the month of November in the year of our Lord one thousand six hundred and twenty-five, and the said act of council made concerning the same of the date aforesaid, in all and sundry points, clauses, and conditions contained therein, and specially, and without prejudice to the foresaid generality, that clause of the same by which it was agreed that the said Provost, Bailies, Councillors, and Community of our said City of Edinburgh and their successors shall have the right, title, and privilege of nomination and presentation of the ministers to serve the cure of all the churches within our said Burgh of Edinburgh in all time coming; and we will, grant, declare, and ordain that this our present ratification of the said heads and articles of agreement, particularly and generally above specified, is and shall be of as much force, strength, efficacy, and effect in all respects as if the same were inserted at length in this our present charter, concerning the non-insertion of which we, for us and our successors, by the tenor

petuum. Quemadmodum nos cum avisamento et consensu predicto tenore presentis carte nostre, damus et concedimus dictis Preposito, Ballivis, Consulibus et Communitati dicte Civitatis nostre Edinburgene, eorumque successoribus, in perpetuum presentationem et nominationem ministrorum ad inserviendum curas, in universis ecclesiis infra dictam Civitatem nostram Edinburgenam, vel hactenus constructuram vel quas in posterum constructe contingent, aliquibus temporibus affuturis, infra dictam urbem, et in unaquaque earum, omnibus temporibus affuturis, unacum jure patronatus dictarum integrarum ecclesiarum, omni tempore futuro, cum plenaria potestate prefatis Preposito, Ballivis, Consulibus et Communitati dicte Civitatis nostre Edinburgene, eorumque successoribus pro perpetuo, qualificatas ad inserviendum curam intra dictas universas ecclesias dicte Civitatis nostre, toties quoties easdem per decessum, deprivationem, demissionem seu quocunque alio modo vacare contigerit, nominandi et presentandi.

. . . .

Igitur proque multifariis bonis servitiis, nobis per dictam Civitatem nostram Edinburgenam, temporibus retroactis, prestitis, et ad incitandum ipsos in eodem, omnibus temporibus affuturis continuari, nos ex

of our present charter, dispense for ever. Moreover we, with advice and consent foresaid, by the tenor of our present charter give and grant to the said Provost, Bailies, Councillors and Community of our said City of Edinburgh, and their successors for ever, the presentation and nomination of ministers to serve the cure in all the churches within our said City of Edinburgh, either already built, or which may hereafter happen to be built in all time coming, within the said City, and in each one of them in all times coming, together with the right of patronage of the whole of the said churches in all time coming, with full power to the foresaid Provost, Bailies, Councillors and Community of our said City of Edinburgh and their successors for ever, to nominate and present qualified persons to serve the cure within the whole of the said churches of our said City as often as they shall happen to become vacant by death, deprivation, demission, or in any other manner.

. . . .

Therefore and for the many good services rendered to us in times past by our said City of Edinburgh, and for inciting them to continue in the same in all time to

nostra scientia et proprio motu, cum avisamento et consensu predicto, de novo dedimus, concessimus, disposuimus, et hac presenti carta nostra confirmavimus, tenoreque ejusdem, de novo, damus concedimus et disponimus ac pro nobis et successoribus nostris, pro perpetuo confirmamus prefatis Preposito, Ballivis, Consulibus et Communitati dicte Civitatis nostre Edinburgene, et eorum successoribus, Totam et integram dictam Civitatem nostram de Edinburgh, menia, fossas, portus, stratas, etc.

. . . . . .

Ac univimus ereximus et incorporavimus, tenoreque presentis carte nostre unimus, erigimus et incorporamus dictas integras terras, portus, receptacula, custumas, propugnacula, et alia predicta ad dictam Civitatem nostram Edinburgenam, ac erigimus eadem in Civitatem regalem, cum integris libertatibus, privilegiis et immunitatibus ad Civitatem seu Burgum Regalem pertinentibus.

. . . . . .

REDDENDO inde annuatim, dicti Prepositus, Ballivi, Consules et Communitas Burgi dicti nostri de Edinburgh, eorumque successores, nobis et successoribus nostris, pro dicta Civitate nostra de Edinburgh, portubus et receptaculis de Leith et Newhevin, cum universis privilegiis

come, we of our knowledge and free will, with advice and consent foresaid, of new have given, granted, disponed, and by this our present charter confirmed, and by the tenor of the same of new give, grant and dispone, and for us and our successors for ever confirm, to the foresaid Provost, Bailies, Councillors and Community of our said City of Edinburgh and their successors, all and whole our said City of Edinburgh, walls, ditches, ports, streets, etc.

. . . .

And we have united, erected and incorporated, and by the tenor of our present charter, unite, erect, and incorporate the said whole lands, ports, havens, customs, bulwarks and others foresaid to our said City of Edinburgh, and have erected the same into a royal City, with all liberties, privileges, and immunities belonging to a City or Burgh Royal.

. . . .

PAYING therefor yearly, the said Provost, Bailies, Councillors and Community of our said Burgh of Edinburgh and their successors, to us and our successors, for our said City of Edinburgh, ports and havens of Leith and

libertatibus, jurisdictionibus, officiis, aliisque supraspecificatis dicte Civitati nostre annexatis, summam quinquaginta duarum mercarum monete sterlingorum, tanquam pro antiquo censu burgali contento in dicto infeofament dicti Burgi, dicte urbi concesso per Regem Robert Bruce, ad duos anni terminos festa viz. Pentecostes et Sancti Martini in hyeme, per equales portiones; ac etiam administrationem justicie in dictis officiis Vicecomitis et Coronatoris ac Justiciarii pacis cum servitiis burgi solitis et consuetis. Ac pro dictis decem acris terrarum de Hierigis, et pro dicto presentatione et nominatione ministrorum curam intra universas ecclesias intra dictam Civitatem nostram constructas et construendas inserviturorum, eidem ut dictum est, unitas et annexatas, duos denarios nomine albefirme, si petatur tantum.

. . . . . .

In cujus rei testimonium huic presenti carte confirmationis, magnum sigillum nostrum, apponi precepimus. Testibus, reverendissimo in Christo patre et predilecto nostro consiliario Joanne miseratione divina Sancti Andree Archiepiscopo, primate et metropolitano dicti regni nostri Scotie, et nostro cancellario; predilectis nostris consanguineis et con-

Newhaven, with the whole privileges, liberties, jurisdictions, offices, and others above specified annexed to our said City, the sum of fifty-two merks money of sterlings, as for the ancient burgh maill contained in the said infeftment of the said Burgh, granted to the said City by King Robert Bruce, at two terms of the year, to wit, Whitsunday and Martinmas in winter, by equal portions; and also the administration of justice in the said offices of Sheriff and Coroner and Justice of the Peace with the services of burgh used and wont; and for the said ten acres of the lands of Hierigs, and for the said presentation and nomination of ministers to serve the cure within all the churches built and to be built in our said City united and annexed to the same as said is, two pennies in name of blanch farm if asked only.

. . . .

In witness whereof we have commanded our great seal to be appended to this our present charter of confirmation. Witnesses, the most reverend father in Christ and our well beloved councillor John, by divine mercy Archbishop of St Andrews, primate and metropolitan of our said kingdom of Scotland, and our chancellor;

siliariis, Jacobo marchione de Hamiltoune comite Arranie et Cantabrigie domino Aven et Innerdaill, Thoma comite de Hadingtoun domino Binning et Byris, nostri secreti sigilli custode, Willielmo comite de Sterling vicecomite de Cannada, domino Alexander de Tullibodie, nostro secretario, dilectis nostris familiaribus consiliariis, dominis Joanne Hay de Barro nostrorum rotulorum registri ac consilii clerico, Joanne Hamiltoun de Orbieston nostre justiciare clerico, et Joanne Scott de Scottistarvet, nostre cancellarie directore, militibus. Apud Newmercatt vigesimo tertio die mensis Octobris anno Domini millesimo sexcentesimo trigesimo sexto, et anno regni nostri duodecimo.

our well beloved cousins and councillors, James marquis of Hamilton, earl of Arran and Cambridge, lord Aven and Innerdaill, Thomas earl of Haddington lord Binning and Byres, keeper of our privy seal; William earl of Stirling, viscount of Canada, lord Alexander of Tulliebodie, our secretary; our beloved familiar councillors Sirs John Hay of Barro, clerk of our rolls register and council, John Hamilton of Orbieston, our justice clerk, and John Scott of Scottistarvet, director of our chancery, knights. At Newmarket the twenty-third day of the month of October in the year of our Lord one thousand six hundred and thirty-six, and in the twelfth year of our reign.

APPENDIX.

I.

[1579, c. 51.]

Act in fauoris of the Hospitall of Edinburgh.

FORSAMEKLE as oure Soucrane Lordis dearest mothir, with auise of hir secreit counsale, efter hir perfyt aige of tuenty fiue yeris, gaif and grantit to the Provest Baillies Counsale and Communitie of the burgh of Edinburgh, for the sustentatioun of the ministrie and hospitalitie within the samyn, all landis annuellis obitis and daill syluer mailis rentis proventis and dewiteis quhatsumeuir, pertening of befoir to quhatsumeuir benefice alterage or chaiplanrie within the said burgh and fredome thairof, or awand furth of the said burgh or tenementis thairof to quhatsumeuir vther benefice or chaiplanrie, as the said gift in itself at mair lenth beris. And albeit thair be certane chaiplanriis foundat in certane places lyand without the said Burgh vnto the quhilkis the saidis Provest Baillies and Counsale ar vndoubtit patronis, the seruice of the quhilkis chaiplanriis hes altogither ceissit and ceissis throw the abrogatioune of the papisticall superstition within this realme, swa that the fruictis rentis and dewiteis of the saidis chaiplanriis of gude ressone aucht now to be applyit to sum better vse. Quhairfoir oure Soucrane Lord, with auise and consent of the thrie estatis of this present Parliament, grantis and gevis full libertie and fredome to the saidis Provest Baillies Counsale and Communitie and thair successouris to ressaue and vplift the proffittis and dewiteis of the foirsaidis benefices alterages and chaiplanriis to the quhilkis thay and thair predecessoris were patronis

pleno jure, to be bestowit to the sustentatiounc of the said ministerie and hospitalitie, and to call follow and persew for the saidis fruictis and proffeittis of all yeiris and termes bipast sen the samyn vaikit, and in tyme cuming for euir, with sic priuilege for recouering thairof as ony ecclesiasticall persone hes for recovering of the fruictis and rentis of thair benefices. Prouiding alwayis that thai be comptable yeirlie to the chekker, and gif thair be ony benefices of cure that they dispone the same to qualifiit personis. And mairover becaus thair ar diuerse personis godlie and zelouslie mouit, petying the miserable estate of the puyr, and delyting in that gude werk of erectioun of ane hospitall within the said burgh, myndit to supplie the said hospitall with thair almous and support of annuell rentis landis and tenementis liand within the said burgh, to be annexit thairto for the intertenyment of the puyr waik aiget and seik personis to be sustenit thairin. Thairfoir our said Souerane Lord, with auise of his saidis thrie estaittis, gevis and grantis licence to all personis quha may be movit to support the said hospitall in landis tenementis or annuelrentis liand within the said burgh sa to do. And that this act of Parliament salbe to the effect foirsaid sufficient in all respectis to all giftis and donationis maid or to be maid of landis or annuelrentis liand within the said burgh to the said hospitall, and salbe of als greit strenth as gif particuler confirmationis wer gevin vpoune euery ane of the saidis donationis as mortifiit to the said hospitall in manner foirsaid, because the samyn can nawise hurt oure said Souerane Lordis proffeitt quha hes na yeirlie proffeit furth of the said burgh bot hes borrow maillis and seruice of burgh.

II.

[1587, c. 8.]

ANNEXATIOUN of the temporalities of benefices to the Crown.

OURE SOUERANE LORD and his thrie estaitis of parliament perfitely vnderstanding the greitest pairt of his proper rent to haif bene gevin and disponit of auld to abbayis monasteries and vtheris personis of clergy, quhairby the crown hes bene sa greitly hurte that thairefter his maist noble progenitouris had not sufficient meanis to beir furth the honour of thair estait as thai had befoir, quhilk hes bred sindrie inconvenientis within this realme, And seeing the causes of the dissolutioun of the patrimonie of the croun to the kirk efter the trewth knawin ar fund nathir necessar nor proffitable, and that be mony occasionis throw a lang procese of tyme the derth has sa gritlie increscit, not onlie in this realme bot in all cuntries, that the princes chairges ar not able to be vphaldin be that part of the patrimonie quhilk now restis in his handis; And his Hienes for the grite luif and favour quhilk he beiris to his subiectis being nawayes myndit to greve thame with importable taxationis specialie for his royall [supporte], It is fund maist meit and expedient that he salhaue recourse to his awin patrimonie, disponit of befoir (the cause of the dispositioun now ceissing), as ane help maist honourable in respecte of him selff and leist grevous to his people and subiectis, AND THAIRFOIR our said Souerane Lord and his saidis thrie estaitis of parliament be the force of this present act haif vnit annext and incorporat, and vnitis and annexis and incorporatis to the crown of this realme to remane thairwith as annext and as it wer propirtie thairof in all tyme cuming and with our said Souerane Lord and his successouris for evir, All and sindrie landis, lordschippis, baronies, castellis, touris, fortalices, mansionis, maner places, milnis, multuris, woddis, schawes, parkis, fischeings, townis, willages, burrowis in regalitie and baronie, annuelrentis, tenementis, reuersionis, custumes greit and small, fewfermes, tennentis, tennendries, and seruice of frie

tennentis, and all and sindrie vtheris commodities proffittis and emolumentis quhatsumeuir, alsweill to burgh as to land (except as heirefter salbe exceptit in this present act) quhilkis at the day and dait of thir presentis, viz., the tuentie nyne day of Julii the yeir of God J^{m} v^{c} fourescoir sevin yeiris pertenis to quhatsumeuir archibischope, bischop, abbot, priour, prioresse quhatsumeuir vther prelate vther ecclesiasticall or beneficit persoun of quhatsumeuir estait, degrie, hie or law, And at the day and dait of thir presentis pertenis to quhatsumevir abbay, convent, closter, quhatsumeuir ordour of freris or nvnis, monkis or channonis, howsoeuir thai be nameit, and to quhatsumeuir college kirk foundit for chantorie and singing, or to quhatsumeuir prebendarie or chaiplanrie quhaireuir they ly or be situate within this realme and dominioun thairof, And sielike All and sindrie commoun landis bruikit be chaptouris of cathedrall kirkis and chantorie colleges as commoun, and quhairof the saidis chaptouris haif bene in possessioun befoir in commountie, to be in all tymes heireftir takin haldin and reput as it wer the propertie and patrimonie of the croun, To remane thairwith in all tyme cuming efter the forme, tenour, and ordour of the act of annexatioun maid in the tyme of oure Souerane Lordis maist noble predicessour King James the Secund, and according to all clausis, conditionis, and circumstances thairof, quhilk in all pointis is haldin for expressit in this present act, And als it is statute and ordainit that the executioun of this act in levying of the proffittis sall begin and tak effect at the terme of Mertimes nixtocum, that our said Souerane Lord may ressaue the rentis and dewties of the said terme and sua furth to continew in tyme cuming.

. . . .

FURTHER oure said Souerane Lord and his said thrie estaittis of Parliament hes declairit and be the tennour heirof declairis, decernis, and ordanis that the landis, lordschippis, and baroneis vnderwrittin, erectit be his Hienes in temporall lordschippis and baronies befoir the dait of this present act, quhilk is the tuentie nine day of Julii the yeir of God J^{m} v^{c} fourscoir seven yeiris, ar not nor sall not be comprehendit in the said annexatioun, excluiding the samin allutterlie thairfra, To remane with the personis to quhome thai wer first disponit efter the forme and tenour of thair infeftmentis maid to thame thairof, thay are to say . . .

The baronies of Brochtoun and Kerse, the burgh of the Cannogait, and ane pairt of the toun of Leyth . . . quhilkis landis and lordschippis are dispouit to diuerse personis as thair particulare infeftmentis beiris.

. . . .

In lykmaner our Soueraue Lord with auise of his saidis thrie estaittis ratifies and apprevis the gift and dispositioun maid be his Hienes to John Bothuell sone lauchfull to Adame bischope of Orknay of the abbacie of Halyruidhouse and haill fructis thairof, with the haill fredomes, jurisdictionis, liberties, rentis, proventis and emolumentis of the samin, with the haill reservationis, claussis, conditionis and circumstances of the same, efter the tennour thairof, as the same of the daitt the tent day of December the yeir of God j^{m} v^{c} lxxxij yeiris beires. . . . .
It is alwayes vnderstand, lyk as our said Soucrane Lord and his thrie estaittis declairis, that vnder the said annexatioun or ony claus heirin specifiet the teyndscheves and vtheris teyndis of quhatsumeuer landis within this realme pertening to ony personnage or vicarage, ar not nor sall not be comprehendit, except quhair the teynd and stok is sett togidder as is heirefter declairit, bot that the samin sall remane with the present possessouris having richt thairto and quha salhave richt to the samyn heirefter, notwithstanding the said annexatioun, or any generall claussis thairin specifeit, quhilkis sall nawayes be extendit to the saidis teyndis mair nor gif the said annexatioun had never been maid. . . . . .
Reservand alwayes and exceptand to all archibischoppis, bischoppis, abbottis, priouris, prioresses, commendatairis and vtheris possessouris of greit benefices of the estait of prelattis and quhilkis befoir had or hes voit in parliament thair principall castles, fortalices, houssis, and mansionis, with the biggingis and yairdis thairof as thai ly and ar situat within the precinctis and clausuris of thair places, quhilkis sall remane with thame and thair successouris heirefter for thair residence and habitatioun notwithstanding the said annexatioun quhilk sall nawayes include the samin. . . . . Provyding alwayes they keip and mantene the samin in the estate quhairin they presentlie ar. Exceptand alsua furth of the said annexatioun all and quhatsumeuer mansionis of personnages and vicarages annext to parroche kirkis with four aikeris of the gleib maist ewest to the kirk and commodious for the

minister serving the cuir thairof for his better residence thairat, quhilkis sall not be nor ar comprehendit in the said annexatioun, but sall remane with the minister, person, or vicare, or vther quha salbe prouidit thairto for serving of the cuir, according to the actis of Parliament maid thair-anent of befoir, EXCEPTAND in lykmaner all and sindrie landis proffittis, tenementis, annualrentis teyndscheves and vtheris emolumentis and proffittis quhatsumeuir gevin, grantit and disponit for intertenement of maisteris and studentis in colleges erectit for exercise of lerning, and for grammer scullis, and for sustentatioun of ministeris makand thair residence in burrowis quhair thair is na vther stipend appointit to thame, AND SICLIKE exceptand and reservand all landis, teyndis, proffittis, annuelrentis and commodities quhatsumeuer grantit befoir the dait heirof be our Souerane Lord, or quhatsumeuer his Hienes predicessouris, or be quhatsumeuer vtheris personis to ony hospitale or masondieu within this realme, and that in favouris of the puir and neidy, Provyding that the same be not disponit nor applyit to ony vther vse. AND FINALIE except-and and excluidand furth of the said annexatioun all landis, baronies, tenementis, annuelrentis and vtheris commodities quhatsumeuer quhilkis pertenit of befoir to quhatsumeuer benefice greit or small being of laic patronages, to the quhilkis the said annexatioun sall not be extendit nor comprehend the same, to the effect that nane of the saidis laic patronis be hurte or damnefiet thairby.

. . . .

THE QUHILK DAY our Souerane Lord sittand in jugement in plane parliament be his declaratioun maid in presence of his thrie estaittis saulfit to him selff liberty and priuilege to except and reserue furth of the actis of the annexatioun of the temporalities of benefices to the croun, of the satisfactioun of the temporalities annext, of his Hienes reuocatioun generall, and of the ratificatioun of the restitutioun, pacificatioun, and abolitioun, all sic personis caussis and materis, and with sic prouisionis, limitationis, and restrictionis as to his Majestie sall seme expedient, quhilkis his Hienes ordanis and commandis his Clerk of Register to insert and incorporat within the bodies of the saidis actis, the same exceptionis and reseruationis being delyuerit to him subscriuit be his Hienes and his chancellair befoir the fyftene day of August nixtocum.

III.

[1587, c. 14.

The Kingis Maiesties generall Revocatioun.

WE JAMES be the grace of God King of Scottis being now of perfite aige of tuentie ane yeiris compleit, and knawing the remeid competent to ws be the commoun law and lawes of our realme, in revocatioun of all and sindrie alienationis, donationis, venditionis, or vtheris dispositionis quhatsumever maid be ws in our minoritie and lesse aige, or be our predicessouris in thair tymes in hurt and detriment of our croun, our saull and conscience, aganis all lawis of our realme, and thairin following the example of our maist noble progenitouris in thair generall reuocationis; and being laitlie past our said perfite age of xxi yeiris, and ane large space within our aige of xxv yeiris, during the quhilk the remeid of our reuocation is competent to ws, We mak our generall reuocatioun in maner following.

. . . .

ITEM we revoik all infeftmentis, giftis and dispositionis quhatsumeuer sett, gevin and grantit be ws in our minoritie to quhatsumeuer persoun or personis in fie fewferme or lyfrent of quhatsumeuir hospitallis maisondewis, landis, or rentis appertening thairto in hurt and preiudice of our conscience, to the end that the saidis hospitallis may be reduceit to thair first institutioun for vphalding of the puir; Prouiding alwayes that the rentis of the hospitall of the Trinitie College besyde the burgh of Edinburgh, quhilk is now decayit, assignit and gevin to the new hospitall erectit to the Prouest, Baillies, and Counsall of the burgh of Edinburgh be nawayes comprehendit vnder this present reuocatioun.

. . . .

ITEM we reuoik all and quhatsumeuer infeftmentis maid be ws in our minoritie our governouris and regentis in our name of ony kirkis landis, freris landis, nvnis landis, or commoun landis, quhilkis onywayes fell and become in our handes as our propertie, except the infeftmentis maid be

our vmquhile derrest moder and ws for erectioun and sustentatioun of hospitallis and ministeris within burrowis quhair thair is na assignatioun nor stipend allowit furth of the thridis of benefices for sustentatioun of the ministeris thairof.

—

IV.

[1592, c. 82.]

Ratificatioun of the landis and annuellis mortifiet to the Ministerie and Hospitall of Edinburgh.

OURE SOUERANE LORD now efter his perfite aige of tuentie fyve yeris compleit, with auise of his estaitis in parliament, Ratifies and Apprevis the donationis and mortificationis maid be his Hienes vmquhile darrest mother in hir perfite aige, and be his Hienes self, at dyuerse tymes, of the landis, benefices and rentis dotit for sustentatioun of the ministerie within the burgh of Edinburgh, and for interteneying of the hospitallis thairof, and speciallie of all annuellis landis and tenementis lyand within the fredome of the said burgh, foundat to quhatsumeuir benefice, and of all landis and annuellis lyand outwith the libertie of the said burgh annexit to ony benefice, prebendarie or religious place, situat within the fredome of the said burgh. And oure said Souerane Lord for his pitifull zeale, quhilk he hes to the sustentatioun of the hospitallis and ministrie within the said burgh, with aduise of his saidis estaitis of parliament, now eftir his perfite aige of tuentie fyve yeris compleit, hes of new annexit to the communitie of the said burgh and thair successouris, in fauouris of thair ministrie and hospitall, all and haill the saidis landis, tenementis, annuelrentis, proffittis and emolumentis foirsaidis, fewfermes, males and dewties thairof, And surrogattis thame in the full richt of all landis annualrentis and emolumentis situat within the fredome of the said burgh, quhilkis pertenit of befoir to quhatsumeuir bischoip, abbot, priour, or quhatsumeuir ecclesiasticall persoun within this realme, And

ordanis ane new infeftment to be exped thairvpoun for thair securitie, gif it be thocht expedient, And for the said godlie effect his Hienes dissoluis the generall annexatioun in that pairt insafar as the samyn may appeir to be extendit to ony of the premissis, or to the annexatioun maid of befoir in fauouris of the said college and hospitall of the kirk of Dumbar[nie], quhairof the kirk of Potie and Mouereiff ar pendicles, lyand within the sherefdome of Perth; the personage of Curry, and the ane half of the vicarage thairof, pertening to the Archedeine of Lowthiane; the landis, annuelrentis, houssis, yardis and biggingis of the Trinitie College situat within the said burgh of Edinburgh, alsweill pertening to the provest as to the prebendaris thairof, and commoun landis and annuelrentis of the same, quhilk annexatioun his Hienes with aduise of his saidis estaitis in parliament ratifies and apprevis, as als his Maiestie, with auise of his saidis estaitis, off new annexis the vther half of the vicarage of Curry, to the quhilk na persoun is prouydit, and the haill vicarage of the said kirk of Dumbarnie quhilk alsua vaikis be depriuatioun of   last possessour of the same, To remane with the Provest, Ballies, Counsall, and Commvnitie of the said burgh and thair successouris in tyme cuming for sustentatioun of thair said ministrie and hospitall. And oure said Souerane Lord and estatis foirsaidis decernis and declaris that nane of thir particularis befoir writtin dispounit of befoir, and newlie annexit for sustentatioun of the said ministrie and hospitall, wer, ar, or salbe euir comprehendit in the generall annexatioun of the ecclesiasticall landis and rentis to the croun, bot wer, ar, and sall be exceptit thairfra lyk as his Maiestie and estaitis foirsaidis off new exceptis the samyn nocht onlie fra the said annexatioun, bot fra all his Hienes reuocationis maid in tyme bypast, or maid in this present parliament, And declaris alsua that the saidis Provest, Ballies, Counsaill and Commvnitie and thair successouris in all tymes cuming, hes and salhaue sic full richt of propirtie and superioritie of the foirsaidis landis annuelrentis and revenewis, tennentis and tenandries, and seruice of frie tennentis thairof, as haid the bischoipis, abbottis, prioris, freiris, monkis, nvnis, chaiplanis, and prebendaris to quhome the saidis landis and annuelrentis pertenit ofbefoir, nochtwithstanding ony act or constitutioun preceding the dait heirof.

V.

[1593, c. 41.]

CONFIRMATIOUN to the Burgh of Edinburgh of thair annuellis.

OURE SOUERANE LORD with avise of his estaitis in parliament hes ratifiet and apprevit, and by the tennour of this present act ratifies and apprevis the act maid ofbefoir in the parliament haldin at Edinburgh the fyft day of Junii the yeir of God J^m v^c fourscoir tuelf yeiris, in fauour of the Provest, Ballies, Counsaill and Communitie of the burgh of Edinburgh, be quhilk oure said Souerane Lord and his estaitis foirsaidis than ratifiet and apprevit the donationis and mortificationis maid be his Hienes vmquhile darrest mother in hir perfite aige, and be his Hienes self at dyuerse tymes sen his Maiesties coronatioun, off all landis, tenementis, annuelrentis, vtheris proffeittis and commodities quhatsumeuir mentionat in the said act, quhilkis wer gevin and doitit for sustentatioun of the ministrie, hospitallis, and college of the said burgh, as at mair lenth is contenit in the said act in all pointis, claussis, articles, and circumstances thairof, quhilkis ar haldin for expressit in this present act. Attour our said Souerane Lord and his saidis estaitis revoikis, retreittis, and rescindis all and sindrie infeftmentis, giftes and dispositionis maid be our said Souerane Lord to quhatsumeuir persoun or personis, off the saidis landis, tenementis, annuelrentis, vthers dewties and commodities quhatsumeuir sen the daitis respectiue of the infeftmentis, giftes and dispositionis maid thairof, to the saidis Provest, Ballies, Counsaill and Comwnitie of the said burgh of Edinburgh for sustentatioun of the said ministrie, hospitallis and college, And declaris the same, with all that followit thairvpoun, to be in all tymes cuming, and to haue bene in all tymes bigane, null and of nane availl force nor effect, And that the personis obtenaris of the saidis infeftmentis, giftis, and dispositionis sall neuir be hard to move actioun nor quarrell thairvpoun, nor found ony exceptioun or defence be vertew thairof aganis the

saidis Provest, Baillies, Counsaill and Commwnitie, and thair successouris, to the effect that thai in all tymes heireftir, without ony impediment or obstacle, may peceablie bruik the saidis landis, tenementis, annuelrentis, vtheris emolumentis and proffeittis, to the vse quhairto thai wer gevin ofbefoir as said is.

VI.

[1606, c. 31.]

Act in fauouris of the Burgh of Edinburgh.

OURE SOUERANE LORD and estaittis of parliament calling to mynd the great charges and expenssis quhairwith the burgh of Edinburgh, burgessis and inhabitantis thairof ar burdenit in sustening of the Ministrie of the said burgh, thair being na pairt of the thriddis of beneficcs assignit to thame for thair sustentatioun, and intertenying of the colleges, hospitallis, and pure of the samin burgh; and that for releif of ane pairt of the saidis charges and expensis his Hienes vmquhile darrest mother eftir hir perfyte age and oure said Souerane at dyuers tymes gaif, dotit mortefeit and disponit to the Provest, Bailleis, Counsaill, and Communitie of the said burgh for sustentatioun of thair said ministrie, and intertenying of thair college, hospitallis and pure, sindrie landis annualrentis, tenementis, provestreis, alterages, benefices, and vtheris fruittis and rentis, generalie and specialie comprehendit and exprest in sindrie giftis, mortificationis, infeftmentis, actis of parliament, and vtheris richtis and securiteis maid to the saidis Provest, Bailleis, Counsaill and Communitie for sustening and intertenying of thair said ministrie, college, and hospitallis; and being cairfull that the foirsaidis godlie necessar and profitable workis decay nocht for laik of expenssis, oure said Souerane Lord, with advyse of the estaittis of this present parliament, hes ratefeit approvin and confermit, and be the tenour heirof, for his Hienes and his successouris, ratefeis, appreves, and perpetualie confermes all and sindrie giftis, mortificationis, infeftmentis, vtheris securiteis and richtis quhatsumeuir maid, gevin, grantit and dis-

ponit be our said Soueranc Lord his vmquhile darrest mother, or be his Hienes self, at ony tyme befoir the date heirof, to the saidis Provest, Bailleis, Counsaill and Communitie of the said burgh of Edinburgh and thair successouris, of quhatsumeuir landis, annualrentis, tenementis, teyndis, provestries, prebendareis, alterages or vtheris benefices, teyndis, rentis, and emolumentis, Togidder with all and sindrie actis of parliament maid in fauours of the said burgh, Provest, Bailleis, Counsaill and Communitie thairof, ministrie, college and hospitallis within the samin, in all and sindrie heidis, pointis, articlis, claussis, circumstances and conditionis thairof, Lyk as our said Souerane Lord with advyse foirsaid ffindis decernis and declaris the foirsaidis giftis, mortificationis, infeftmentis, actis of parliament, and vtheris richtis and securiteis maid gevin and grantit to the saidis Provest, Bailleis, Counsaill and Communitie of the said burgh and thair successours, and in thair fauours, to be gude, valide, lauchfull and sufficient richtis for thame and thair successouris for bruiking and joising of all and sindrie landis, tenementis, annualrentis teyndis, benefices, provestreis, prebendareis, chaplanreis, alterages, fruittis, rentis, and emolumentis quhatsumeuir, alsweill generally as speciallie comprehendit and contenit in the foirsaidis giftis, mortificationis, actis of parliament, or ony of thame, And that the samin sall remane and abyd in thair awin full strenth force and effect, and nawayes to be tane away hurte nor prejudgeit in ony sorte, nochtwithstanding quhatsumeuir act or statute made in this present parliament, with speciall provisioun that thair be ane ressonable and sufficient stipend modefeit be George archibischop of Sanctandrous, Master Johne Prestoun of Pennyeuk, collectour, and Sir John Skene of Curryhill, knycht, clerk of register, or be ony twa of thame, the said archibischop of Sanctandrous being alwayes ane, to the present minister at Currie for him selff and to his successouris ministeris thairat, to be payit thankfullie to thame yeirlie in all tyme cumming be the towne of Edinburgh, The quhilkis thrie personcs being all present in parliament acceptit the modificatioun of the said ministeris stipend at the said kirk of Curry in and vpoun thame, to be payit to the present minister and to his successouris ministeris thairat yeirlie in all tyme cumming abonespecefeit.

VII.

WARRANT for an Act of Parliament superscribed by King James the Sixth, ratifying all previous Grants of Kirk livings to the Burgh of Edinburgh. 22d October 1612.

WARRANT for an Act of Parliament superscribed by King James the Sixth, whereby His Majesty and estates of Parliament ratify and approve the infeftment granted by His Majesty under the Great Seal, with advice and consent of the Lords of Privy Council, to and in favour of the Provost, Bailies, Council and Community of Edinburgh, and their successors, for entertainment of the Ministry serving the cure at the kirks of the said burgh, and of the Masters and Regents of the College thereof, Schools and Poor within the samin burgh, of All and Haill the benefice of Trinity College situate beside the said burgh at the foot of Leith Wynd, and of certain other benefices, kirks, lands, teinds, annualrents, tenements, profits, emoluments and others at length mentioned and set down in the said infeftment. To be holden of his Majesty and his successors in free alms for ever, of the date at Edinburgh (but it ought to be at Beavoir Castle, being so named in the Signature), the 7th of August last bypast, viz. 1612, in the haill heads, articles and clauses thereof. And His Majesty and estates foresaid willed and granted, statuted and ordained that the foresaid infeftment, and this ratification thereof, were and should be good valid and effectual rights and securities to the said magistrates and their successors for bruiking and enjoying the lands, benefices, teinds, and all other privileges, liberties, and commodities contained in the said infeftment in all time coming; ordaining the Lord Clerk of Register and his deputes to extend an Act of Parliament thereupon, and to insert and engross the infeftment before specified in more ample form thereintil. This bears date 22d October 1612, and underneath is written thus:—"This containeth a ratification of all church lands granted at any time heretofore to the burgh of Edinburgh. (Signed) Io. DOUGLASSE," and is superscribed by the King as said is.

VIII.

[1621, c. 79.]

RATIFICATIOUN of diuers infeftmentis grantit to the toun of Edinburgh for sustentatioun of Colledge, Ministeris and Hospitallis.

OURE SOUERANE LORD and estaittis of this present parliament considdering that his Majestie with auise of the Lordis of his Hienes secreit counsell, ffor the gude and thankfull seruice done to his Hienes and his most noble progenitoures be the Provest, Bailyeis, Counsell and Communitie of the burghe of Edinburgh, and for the gryit zeale quhilk his Majestie cariet to see the youthe tranit vp in learning and vertew, gave, grantit and disponit to the saidis Provest, Bailyeis, Counsell and Communitie of the said burgh of Edinburgh, and thair successoures, licience and libertie to erect ane Colledge, and builde and repair sufficient housses and places for receptione of proffessoures of humane letteris and toungis, of philosophie, theologie, medicine, the lawes, and of all vther liberall sciences, and to erect and chuis sufficient professoures for teiching the saidis professiounes; and to that effect gave, grantit and disponit to thame and thair successoures the provestrie of the Kirk a Feild with the landis, tenementis, fruittis, possessiounes, rentis, dewties and pertinentis of the same, as at mair lenth is contenit in the chartour and infeftment grantit to thame thairupoune, vnder his Hienes gryit seale, of the date at Stirling the fourtene day of Apryill Jm vc fourscore tua yeiris. Lyikas conforme thairto the saidis Provest, Bailyeis and Counsell of Edinburgh biggit, edifiet, and repairit ane gryit ludgeing, togidder with the manse and house of the said provestrie of the said Kirk of Feild to the vse of ane Colledge for professioun of philosophie, theologie, and humanitie, quhilk Colledge hes continewallie sensyne flurischit now be the space of threttie fyve yeiris. And his Majestie schortlie efter the bigging and edifeing of the said Colledge, viz. in the moneth of 1584, considdering the saidis Provest, Bailyeis, Counsell and Communitie of the said burgh of Edinburgh had bestowit gryit chargis and

expensses in erectioun of the said Colledge, building of housses, and had doted gryit sowmes of money for intertenement off professouris of humanitie, philosophie and theologie within the samen, ffor instructing of the youthe thairin, gave, grantit and disponit to the saidis Provest, Bailyeis, Counsell and Communitie of the said burgh to the vse of the said Colledge, and for sustentatioune off the rectour and regentis within the samen, All and Haill the Archedenerie of Louthiane, conteaning thairin the personage of Curryc with manse, gleib, kirklandis, teindis, fructis, rentis, proffittes and dewties of the samen, as at mair lenth is contenit in the chartour grantit thairvpoun, off the dait the fourt day off Apryll 1584 yeiris, vnder his Hienes gryte seale. Lyik as his Majestie considdering the great chargis and expenssis debursit be the saidis Provest, Bailyeis, Counsell and Communitie off the said burgh of Edinburgh in erectioun of ane hospitall vpoun intertenement of thair Ministerie and Colledge foirsaid, gave, grantit, and disponit to the saidis Provest, Bailyeis, Counsell and Communitie of the said burgh of Edinburgh and thair successouris All and Haill the provostry of the Trinitie Colledge, landis, houssis, rentis, kirkis, teyndis, vtheris fructis, rentis, and emolimentis thairto annexit, mair at lenth contenit in the said chartour vnder his Hienes gryit seale off the date the tuentie sext day of Maii 1587 yeiris. Lyik as his Majestie efter his perfyit age of tuentie ane yeiris compleit, be his lettres patentis vnder his gryit seale, ratiefiet, apprevit, and confermit the foirsaidis chartours off the contentis and daittis respectiue foirsaid, as at mair lenth is contenit in the Chartour of Confirmatioun off the samen vnder his Hienes gryit seal off the date the 29 of Julii 1587. Lyik as his Majestie be his new chartour vnder his Hienes gryit seale off new, gave, grantit and disponit to the saidis Provest, Bailyeis, Counsell and Communitie off the said burghe of Edinburgh and thair successours, for intertenement of the said Colledge, Ministrie and Hospitall, All and Haill the foirsaid provestrie of the Trinite Colledge, with vtheris certane benefices, kirklandis, teyndis, annuelrentis, tenementis and vtheris mair at lenth sett doun in the said chartour and infeftment, to be haldin in maner thairin conteaned, quhilk is of the date at Bearvoir Castell the sevint day off August 1612. Lyik as his Majestie considdering that sindrie and dyuerse godlie and weildis-

posed persounes hes doted and mortefied dyuerse and sindrie gryit soumes of money to the vse of the said Colledge, and for sustentatioun of professouris of humanitie, theologie, and certane bursaris within the same. And his Majestie out off his gratious lufe, affectioun, and royall care quhilk he beiris to the grouthe and incres of leirning within this realme, and speciallie within the said burgh of Edinburgh, being his Majesties principall toun and burgh within this his kingdome, being maist willing that the formar giftis and particularis grantit thairto be confermit, and all farder immunities grantit thairto in als ample forme as any vther colledge hes or bruikes within this realme. Thairfore his Majestie with aduyse of the estaittis of this present parliament, ratifies and approvis the foirsaidis infeftmentis grantit be his Majestie vnder his Hienes gryit seale, togidder with the erectioun of the said gryit ludgeing manse and hous of the Kirk of Feild in ane Colledge for professioun of theologie, philosophie and humanitie, togidder with the foirsaidis mortificatiounes maid be his Majestie ather to the vse of the Colledge, or to the vse of the Ministrie and Hospitall of the said burgh, in all and sindrie poyntis, passis, heiddis, articles, conditiounes priuiledgis, immunities, liberties and vtheris circumstances quhatsumeuir thairin contenit, efter the forme and tennour thairoff. Lyik as his Majestie and estaittis foirsaidis statutes and ordanis that this present ratificatioun is and salbe als valide effectuall and sufficient in all respectis as giff the foirsaidis infeftmentis of the dates respectiue abonewrittin were at lenthe and worde be worde ingrossit in this present act. And sielyik his Majestie and estaittis foirsaiddis willes, grantis, statutes and ordanis that the foirsaidis chartours and this present ratificatioun of the samen ar and salbe guide, valide, lauchfull and effectuall rightis and securities to the saidis Provest, Bailyeis, Counsell and Communitie of the said burgh of Edinburgh, and their successours, for bruiking and joising of the landis, benefices, teindis, and for erectioun of the said Colledge, and all vther priuiledgis, liberties, and commodities contenit in the saidis chartours in all tyme cuming, efter the forme and tennour thairoff. Ordaning the Clerk of Register and his deputtis to extend ane act of Parliament heirupoun, and to insert and ingrose thairintill the chartouris abonespecifiet in mair ample forme. Lyik as his Majestie off his princelie and royall

fauour, and for gude seruice done to him be the saidis Provest, Bailyeis, Counsell and Communitie of the said burgh of Edinburgh, and for thair further encouragement in repairing and reedifeing off the said Colledge, and placeing thairin sufficient professoures for teiching of all liberall sciences, ordaning the said Colledge in all tyme to cum to be callit King James Colledge, and als with aduyse of the saidis estaittis hes of new agane gevin, grantit and disponit to thame and their successouris in fauores of the said burgh off Edinburgh, patrones of the said Colledge and off the said Colledge, and of the rectouris, regentis, bursaris, and studentis within the samen, all liberties, fredomes, immunities, and priuiledgis appertening to ane free Colledge, and that in als ample forme and lairge maner as anye colledge hes or bruikis within this his Majesties realme: And gif neid beis ordanis ane new chartour to be exped vnder his Hienes gryit seall for erecting of the said Colledge, with all liberties, priuiledgis, and immunities quhilk anye colledge within this realme bruikis, joises, or to the samen is knawin to appertene. And for thair farder securitie his Majestie and estaittis hes dissoluit and dissoluis the foirsaidis haill giftis and vtheris particularlie abonespecifeit fra his Majesties croun, and fra all actis of Parliament maid thairanent, in sa far as the samen or anye pairt thairoff is or may appeir to have bene annexit thairto in tyme bigane, to the effect particularlie abonespecifeit. And annullis all and quhatsumeuir actis and statutes quhilk may be preiudiciall or dirogatorie to the premisses. And giff neid beis ordanis new giftis to be exped thairvpoun.

IX.

[1661, c. 123.]

Ratification of his Majesties new Charter of Confirmation in favors of the Burgh of Edinburgh.

Our soverane lord with advice and consent of his estates of parliament, ratifies, approves and confirmes ane signatour granted and supersigned be his Majestie at Whitehall the tent day of September 1660 yeers, ordaineing ane charter to be past vnder the Great Seale of Scotland ratificing and approveing, and for his Hienes and his successours perpetually confirmeing in favours of his Majesties antient burgh of Edinburgh, Provost, Baillies, Councill and Communitie thairof and thair successours, all and whatsoever charters, infeftments, confirmations, gifts, grants, donations, mortifications, decreits and sentences, acts of Parliament, acts of Generall Conventions, acts of Secreit Councill and others writts and evidents whatsomever made, granted, or confirmed by any of the Kings or Queens of Scotland, Governours or Regents thairof for the tyme, or by thair Commissioners, or by whatsomever other persone or persones to the forsaid burgh of Edinburgh, or to the Kirks, Colledge, Ministers, and Hospitalls of the said Burgh, of whatsumever forme or maner tenor or contents date or dates the same be of, in the haill heids, articles, clauses and conditions thairof, or to the Magistrats, Councill, Burgesses and Commonalitie of the same burgh of Edinburgh, or in favours of the Ministers, Kirks, Colledges and Hospitalls thairof, together with all erections, liberties, rents, lands, tenements, jurisdictions, superiorities, mortifications, patronages and others whatsoever perteaning and disponed to them, and whairof they or thair predicessours are or have been in possession. And his Majestie willed and declaired that the said generall confirmation is and shall be in all tyme comeing of als great strenth, force, and effect to the said burgh of Edinburgh, and to the Provest, Baillies, Councill and Communitie

thairof and thair successours, and to the Ministers, Colledges, Hospitalls and Poore thairof, as if all and sindrie the saidis infeftments, gifts, dispositions, mortifications, alienations, acts, decreits, confirmations and other securitys wer therin at lenth ingrossed and repeited (Reserve-and to his Maiestie and his successours the dewties, rights and services vsed and wont to have been payed of befor) togther with the forsaid charter appointed to passe vnder the Great Seall vpon the said signatour, in the haill heids, articles, clauses and conditions thairof: Willing, declareing and ordaineing thir presents to be als valeid and sufficient as if the said signatour and charter to follow thairvpon wer alreadie extendit and past vnder the Great Seall, and that the same, togther with all and sindrie infeftments, gifts, dispositions, mortifications and other rights, tytles and securities forsaids therby ratified and approven, wer heirin at lenth word be word insert and ingrossed. And his Maiestie, with advice and consent of his estates of parliament, decernes and ordaines the forsaid signatour and charter to passe thervpon with this present confirmation thairof to be ane good and perfyte right to the saids Provest, Baillies, Councill and Communaltie of Edinburgh and thair successours for brookeing and ioyseing conform to their rights, the haill lands, tenements, liberties, rents, iurisdictions, superiorities, mortifications, patronages, offices and others whatsumever granted be his Majesties dearest father of blessed memorie King Charles the First, or any others his royall predicessors to the said burgh of Edinburgh, or to the Kirks, Colledges, Ministers and Hospitalls thairof. And ordaines his Majesties thesaurers principall and deput and remanent Lords of Exchequer to passe to the said burgh of Edinburgh thair Colledges and Hospitalls the particular infeftments and grants forsaid.

www.ingramcontent.com/pod-product-compliance
Lightning Source LLC
Chambersburg PA
CBHW021705210326
41599CB00013B/1522

* 9 7 8 3 3 3 7 2 3 5 9 7 0 *